THE AMATEUR HORSE BREEDER

(REVISED EDITION)

Also by A.C. Leighton Hardman

STALLION MANAGEMENT

THE AMATEUR HORSE BREEDER

(REVISED EDITION)

Written and illustrated by

A.C. Leighton Hardman

PELHAM BOOKS

First published in Great Britain by

PELHAM BOOKS LTD
52 Bedford Square
London, W.C.1
JUNE 1970
SECOND IMPRESSION AUGUST 1971
THIRD IMPRESSION DECEMBER 1972
FOURTH IMPRESSION NOVEMBER 1973
FIRST PUBLISHED IN THIS REVISED EDITION SEPTEMBER 1974

ISBN 0 7207 0796 X

Printed in Great Britain by
Hollen Street Press Ltd, Slough
and bound by James Burn at Esher, Surrey

To

My Parents and Brother

CONTENTS

ILLUSTRATIONS

DRAWINGS

ACKNOWLEDGEMENTS

I should like to express my gratitude to Leo B. Jeffcott, M.R.C.V.S. for his helpfulness and kindness to me in reading and correcting this book, also to the veterinary surgeons at Reynolds, Leader, Day and Crowhurst, Newmarket for their invaluable advice, especially to R. J. McEnery, M.R.C.V.S. and D. Simpson, M.R.C.V.S. To L. Wilson Esq., of Pegus Ltd. for permission to reproduce the Equilac chart. To Mrs P. J. Leader for putting up with the noise of a typewriter. I am also grateful to Ann and Tina for all their help, especially to Ann for the unlimited loan of her typewriter.

NEWMARKET A.C.L.H.

FOREWORD

By Major-General Sir Evelyn Fanshawe
C.B. C.B.E. D.L.

The Author of this book though young in years has made a very thorough study of her subject and what is more has taken great trouble to get practical experience in breeding horses and ponies.

The subject of horse breeding can be matter for many volumes but this book is a sound and handy guide for the small breeder.

The opening paragraph is very true and puts in a nut shell why most of us want to breed a horse.

This book is written in simple language: the subject matter is thoroughly sound and I recommend it to all breeders and particularly to those who are just starting.

Evelyn Fanshawe

I

YOUR MARE

A foal of your own: something you have bred yourself from your own mare, by a stallion which was entirely your own choice, halter broken and handled by no-one but you – in fact, all your own work; this is the dream and ambition of many horse owners but unfortunately realised by only a few.

Perhaps one of the main reasons why more people do not breed their own foals is just lack of knowledge, as this is a subject outside the realm of everyday horse management but one in which anyone with the know-how can become really interested, as the pride of ownership in a bought animal is nothing compared with that felt by the breeder.

When deciding whether or not to breed a foal from your mare one of the most important considerations is her age:

It is unwise to get a mare in-foal before she is at least three years old, as horses continue to grow and develop until they are about five years old but grow and develop very fast up to three years old and to get a mare in-foal at this critical period in her life, could stunt her for ever, unless she is very well fed.

Some authorities state that the most difficult age to get a mare in foal is at three years, though from my own experience I have not found this to be particularly so. However, mares over 15 years old with increasing age tend to become less fertile but if kept breeding regularly will often go on producing foals into their twenties.

There are on record instances of mares producing their first foals in old age and as long as the mare comes in season regularly and puts up a good egg, there is absolutely no reason why she should not get in-foal, even if she is 20 or more years old and there is no real reason why she should have a difficult foaling. For unlike her human counterpart, the mare can remain fertile until she dies of old age.

An equally important consideration is the conformation and condition of the mare's internal and external genital organs. We will deal first with the external organs, as these can be checked by the owner himself, whereas only a qualified veterinary surgeon can check the internal organs satisfactorily.

First of all look at the mare's udder, make sure that it is free from warts. Next, lift her tail and compare the conformation of her vulva with the diagrams below :

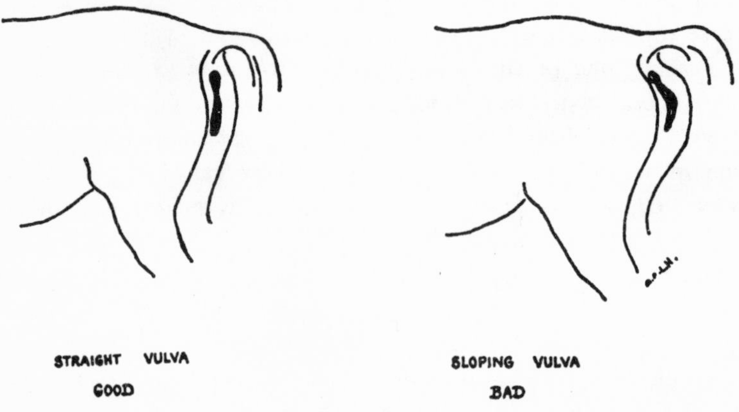

STRAIGHT VULVA
GOOD

SLOPING VULVA
BAD

A mare with a sloping vulva is seldom a good breeding proposition in the natural state. At times most especially when in season, she will be heard to draw air in through her vulva, this air passes down into the uterus where it can set up microbiological infections which in turn can lead to sterility, loss or reabsorbtion of the foetus. This is commonly known as "windsucking" – not to be confused with the vice of the same name!

This malformation of the vulva can, however, be rectified

by a simple operation (known as a Caslick's Operation) carried out by your veterinary surgeon. Using a local anaesthetic he simply cuts down the sides of the vulva and stitches these parts together down to the brim of the pelvis; this prevents the lips of the vulva from hanging loosely open at the top end and enabling air and dung particles to enter the vagina. The stitches are removed in about 10 to 14 days. This operation does not in anyway interfere with the mare's ability to urinate freely nor unless she has been stitched down a long way with subsequent service by the stallion, if this is so the bottom stitch is removed and replaced after service.

If your mare happens to be a grey, turning or already turned to white, on looking under her tail and in the region of her vulva you may notice a collection of small hard black lumps, sometimes known as Melanoma or 'black cancer' – these are tumours which, due to the loss of coat colour with advancing age, become heavily laden with the black pigment – melanin. Generally speaking they are nothing to worry about and very seldom interfere in any way with her foaling. It is however a good idea to have a mare with these lumps examined by a veterinary surgeon before getting her served just to make sure that she will be all right.

A mare should come in-season regularly once every three weeks and stay in season on average from three to nine days. If you do not know whether she comes in-season regularly or indeed if at all, it is probably better that you should have her checked internally by a good horse veterinary surgeon before sending her to stud. Sometimes mares have cystic-ovaries and consequently are in-season all the time, unless these can be cured the mare is very unlikely to breed. If on the other hand, the mare never comes in-season, there is always the possibility that she may already be in-foal, especially if you have only just bought her and aren't entirely familiar with her background! Either way you will be wasting your money and the stud's time by sending the mare to them before you have had her examined.

When you are deciding to get a foal from your mare, make

up your mind quite definitely what type of horse or pony you wish to breed.

Stand back and look at your mare critically, if you can only see good in her ask a knowledgeable friend to help you. If she has any outstandingly bad points, choose a stallion which is particularly good in this respect. Each parent contributes 50% to the genetical make up of the off-spring, so it is not always possible to get rid of any particular characteristic in the first generation though an exceptionally prepotent stallion will help.

Generations of unsound stock, which go lame immediately they are put into work, can and indeed are bred, when unsound stallions and mares are used for breeding. Unsoundness in this case does not include accidental injury, which can never be passed on to the resulting foal but only relates to those hereditary diseases which are listed on page 24.

It sometimes happens that an unsound mare which is suffering from one of these listed diseases, is offered free to a good home for breeding purposes. To safeguard the soundness of the horse for future generations, these mares should never be bred from, however tempting the offer may seem.

Regarding size: the first foal is usually the smallest and where there is any very great discrepancy in size between the parents, the foal will, at birth, more often follow nearer to the size of the mare than the stallion, this being nature's safeguard for an easy foaling.

When considering Half-bred and pony mares of mixed breeding, better results will be obtained by sending them to pure-bred stallions; when using stallions of mixed parentage, however good, on mares of unknown or mixed breeding, any one of a vast number of characteristics may emerge with disastrous results. A stallion of pure ancestry tends by his many generations of pure breeding to be more prepotent, i.e. able to pass on his own characteristics in preference to those of his mate.

To obtain more size and quality in the first generation there is nothing better than an Arab or Thoroughbred sire. The Arab, due to his many centuries of pure desert breeding is the most prepotent breed of horse today and so the most capable, of obtaining maximum quality, from half-bred mares in the first generation. He is closely followed in this respect by the Thoroughbred, which can impart that extra height as well, which is so essential in hunter breeding; whereas the mare in turn can, if she has substance herself, help to produce the bone, i.e. the circumference of the leg immediately below the knee, measured in inches: a good hunter should have size, quality, and at least 9 inches of bone, an ideal hunter sire should, therefore, have both size, quality and good flat bone.

Before you finally decide to send your mare away to stud, consideration should be given to the amount of accommodation available. You will need double the grazing area, i.e. a minimum of four acres for every mare and off-spring for all the year round grazing and double the loose box space. Initially, there will be little change (1-2 acres is sufficient) but in a matter of about four months, the foal will be eating a considerable quantity of grass each day and by the time it is only six months old, its daily requirements will be almost identical to those of the mare. Therefore, it is unwise to breed from a mare unless you are sure you have the room to deal with *two* horses.

2

CHOICE OF STALLION AND STUD

When you have finally chosen the type of stallion you wish to use, the next thing to do is to decide how much you intend to spend on the stud fee and look around for a suitable stallion.

It is as well to remember that it costs as much to rear a bad foal as it does a good one and ten pounds spent at this stage in travelling expenses or stud fee to a good stallion could make £100 difference to the value of the resulting foal. So don't use a local stallion just because he stands locally, if, on the other hand, your local stallion is also the best, then count yourself lucky and by all means send your mare to him.

There are several ways of finding out which stallions are standing at stud and where they are located. One of the best ways is to write to the breed societies concerned – most keep lists of stallions standing at stud in the British Isles and will gladly send you a copy if you write to them.

Other useful methods of selection are to buy copies of *Horse and Hound* any time from November to May – this magazine carries an extensive range of advertisements of stallions at stud which embrace most breeds and types. Ask your Farrier if he knows of any stallions at stud in your area or buy a local paper, as most of these contain advertisements during the breeding season, of local stallions standing at stud.

Using any or all of these methods, compile a list of stallions in which you are genuinely interested and write to the studs concerned, asking for the following:—

(1) A stud card – this should give you a photograph of the horse and details of his breeding, winnings and winning progeny, if any; also his stud fee.

With regard to the stud fee, this is basically divided into four categories:

(a) Fee without any concessions – this must be paid regardless of whether the mare is in-foal or not.

(b) No foal, free return – this is self-explanatory but the free return must be taken the following year.

(c) No foal, no fee – in this case the mare must be tested barren by October 1st otherwise the fee is due.

(d) No live foal, no fee – the stud fee is usually payable (at the end of the stud season) unless the mare is tested barren but the fee is returnable in full, if the mare does not produce a live foal which survives for 48 hours. Sometimes a stallion is advertised as first a live foal then payment of the stud fee.

The last two concessions usually indicate that the stallion concerned has a high fertility and is well managed, otherwise the owner would not risk the loss of fees involved. Stallions standing at no foal, no fee or with live foal concessions usually have slightly higher stud fees, to compensate for the loss of income from mares which fail to conceive – the national foaling average for Thoroughbred mares is only two foals for every three years at stud. The stallion owner also reserves the right to refuse any mares he considers unsuitable for any reason – this is stated as 'approved mares only'.

However, mares visiting stallions which do not offer any concessions can be covered by insurance (with Lloyds) against loss of stud fee and expenses incurred whilst at stud, providing the stud fee is more than £45. The premium for this is, however, rather high. The average premium being 40-50% with 10% returned in the event of no claim.

(2) Fertility percentage – this is the number of mares got in-

foal in a given year worked out against the total number of mares served by the stallion. Ideally this should be at least 70%.

(3) Ministry of Agriculture Stallion Licence – it is as well to ask if the stallion is licensed, as all stallions other than Shetlands and those registered in the General Stud Book, which stand at public stud and all travelling stallions must have a Ministry Licence, which means that they have been passed by a veterinary surgeon as being of good conformation and free from the following hereditary diseases:

> Cataract; Shivering; Defective genital organs; Roaring and Whistling; Navicular disease; Sidebones; Ringbone (high and low); Bone Spavin; Stringhalt.

It should be noted that any mare used for breeding should also be free from these defects as she would be just as likely as the stallion to pass them on to her foal.

Each year in March some Thoroughbred stallions, which are registered in the General Stud Book and many of which have won races, compete for premiums at the Stallion Show in Newmarket. These premiums are subsidised by an annual grant from the Horse Race Betting Levy Board and this enables the stallions to travel the different districts in England, Wales and Scotland, to which they are located, by the Hunters Improvement and National Light Horse Breeding Society, at a very reduced fee.

These stallions have all passed a rigorous veterinary examination and are therefore guaranteed not to possess any hereditary unsoundness which they could pass on to their off-spring. As already discussed in Chapter One, this is of great importance in breeding horses.

For the small breeder wishing to breed National Hunt horses, Point to Point horses, Hunters or Hacks from their mares, these stallions represent the best value for money anywhere in this country.

Many of these stallions have even produced winners on the flat as well as many well known National Hunt winners such as:

Highland Wedding, Pas Seul, and Midnite Masquerade. Many of the top show hunters, both in-hand and under saddle are also by these horses, as are some very well known Three Day Event horses and some of our top Show-jumpers.

If you do decide to use a Hunters Improvement Society Premium Stallion on your mare, besides having a very low stud fee to pay, your mare will also be eligible to compete at certain shows for a premium, offered to mares which have produced a foal, by a stallion, which was awarded a premium under the Society's scheme in the year of covering. Three premiums of £15 each are usually allotted in each class. In all cases the mare must have her foal at foot and no mare can take more than one premium in any one year, except for the Hunters Improvement Society Summer show and the Royal of England and Royal Highland shows. The winning mare must be certified as being free from hereditary disease by the show's Veterinary Officer, and she must then be entered in the Hunter Stud Book and her owner must also be a member of the Hunters Improvement Society, before payment of the premium is made. The premiums being paid direct to the successful exhibitor by the Hunters Improvement Society.

The Arab Horse Society also runs a brood mare premium scheme, this is open to mares which have been covered by a registered Arab stallion, these premiums differ from those awarded by the Hunters Improvement Society in that they are competed for in the year of covering, but the mare should have a foal at foot or be certified in-foal. These premiums are only awarded at certain shows approved by the Arab Horse Society. The premiums are awarded to mares which in the judges opinion are most likely to produce a good foal when sent to an Arab stallion, the premiums are given as follows, and like the Hunters Improvement Society Scheme only one premium can be won in any year by each mare, except for the Arab Horse Society's Annual Summer show:

Class A—Thoroughbred, Hunter and Hack Type Mares over 14.2 h.h.

Premium

Mares entered in the General, Arab Horse Society's (Anglo-Arab Section) or Prior's Half-bred Stud Books £15

Other Mares (including Part-bred Arabs) £10

Class B—Pony Mares, 14.2 h.h. and under.

Mares entered in the General, Arab Horse Society's (Anglo-Arab Section), Prior's Half-bred or recognised Mountain or Moorland Stud Books *OR* descended from a cross between an Arab and a Mountain or Moorland Pony whose Sire and Dam are entered in a recognised Stud Book, i.e. where entered in the Part-bred Arab Register £10

Other Mares £8

(4) Charges for keep – For Thoroughbred studs these are usually divided into three categories:

viz. Foaling mares,
Maiden mares,
Barren mares.

i.e. Those about to foal or with foals at foot; those which have never been served before and those which have had foals before but which are without foal at the time of going to stud; there is another class of mare known as a barren-maiden: this is a mare which has been served before but which has never had a foal.

Pony studs usually only divide their keep into two catagories:

viz. Foaling mares,
Barren or Maiden mares;

but these are sub-divided according to whether the mare is kept at grass or is stabled.

Keep charges vary enormously and can range from as little as £2.00 per week at grass on a pony stud to about £25.00 per week for a foaling mare on some Thoroughbred studs.

Studs making very low charges for keep should be avoided as nobody will keep your mare for free and most studs like to make a profit, to help off-set their labour expenses, so low charges often mean that your mare will be returned to you much thinner than when she went to stud.

A groom's fee is usually charged and varies from £1.00 to £4.00 per mare.

Any veterinary charges are sent to you direct by the vet. attending the stud. While worming, shoeing and transport expenses are charged as extras.

Mares usually stay at stud for about twelve weeks, so a rough estimate of the total charges likely to be incurred at a pony stud are as follows:

	£p
Stud fee	25.00
Groom's fee	1.00
Keep at £5.00 per week ...	48.00
Blacksmith	2.00
Worming	1.00
	£77.00

These charges are of course correspondingly higher if the mare goes to a Thoroughbred stallion, other than a Hunters Improvement Society Premium Stallion whose fees for Half-bred mares are similar to those set out above: i.e. 12 gns for members or 21 gns for non-members.

Once you have sorted through the stud cards you received, you will have an idea of the stallions in which you are still really interested. The next step to take is to telephone the studs concerned and make an appointment to see the stallions; it is very necessary to telephone the stud manager before you go, as most studs are extremely busy places during the breeding season.

When you visit the stud first look at the stallion in his box, apart from his conformation pay special attention to his temperament, this is most important if you are going to look after and handle the resulting foal yourself, as bad temper or a mean disposition can be inherited. Next ask to see him out and ask the groom to trot the horse up for you – make sure his action is straight and that he moves well: i.e. points his toe. Ask the owner or stud-groom if you can see some of the stallion's progeny and if possible the mares these are out of – this should then give you some idea of the type of foal you can expect from your own mare. If your mare is in-foal, it is as well to ask the stud groom whether the foals are handled and led, otherwise you will be presented with a very big wild foal by the time he is two or more months old and your mare has been tested in-foal again.

While you are looking round the stud check the condition of the animals, type of fencing, state of the paddocks and the loose-boxes, i.e. if they are clean, tidy and in good repair and run to the same or a higher standard than your own stable at home.

It is probably very difficult to see your own yard as other people see it but your first impression of any stud should be one that is clean and well swept, with the boxes in good structural repair, painted or creosoted and obviously mucked out every day. As far as your own or any stud is concerned, there should be no broken windows and all windows should have guards over them, there should be no nails sticking out in either the boxes or paddock fencing and no holes in the boxes especially low down, all doors must reach ground level as a horse on lying down, could put a leg through a gap under the bottom of a door and get badly trapped. The water bucket or water bowl in the boxes must be cleaned out every day as should the field troughs in the paddocks. It goes without saying that there should be no barbed wire on any well run stud and there must be no unguarded open ditches into which foals or even older horses could fall.

The mares with foals at foot should be kept in separate paddocks from those without foals, to prevent accidents.

These points are very important as of course, no stud accepts the responsibility of your mare from the point of view of accidents or disease, although naturally every care is taken of her, but the ultimate responsibility rests with the mare owner. Nevertheless it is equally important that your own stud should be maintained at an adequately high standard – otherwise avoidable accidents will occur, ending in total losses or reduced values of stock from injury, as young stock are notoriously more prone to accidents than older horses.

When you have finished looking at the stallion and have been shown round the stud, it is always very much appreciated if you give the stallion man or stud groom a tip, especially if you intend to send your mare to the horse.

When you have visited all the studs on your list you will be in a position to decide exactly which stallion you intend to use. Write to the stud of your choice and ask them to send you a nomination form. This is a form signed by every mare owner, when they have booked a nomination to a particular stallion, in which they agree to send a mare to that stallion at a fee stated on the form, these forms are usually in duplicate – one section being retained by the mare owner, and the other by the stud. Most studs also require details of the mare to be sent, and it is most helpful if you can supply the stud with as much information as possible. It is as well to do this early in the year, as most of the best stallions get booked up quickly – an adult stallion normally only taking forty mares each season.

.................19......

Dear

I shall be much obliged if you will kindly fill up and return to me the annexed form confirming one nomination taken to

....

for season 19........... at ....

Yours faithfully,

Mares must NOT be sent shod.

Woodfield House Stud.

(Every precaution will be taken against accidents or disease but no liability can be accepted therefore.)

No. ....

....

....

....

..........................19......

Dear ....

I agree to take one nomination to

....

for season 19........... on the following terms :

Stud fee .... Groom's fee ....

Keep of Mare ....

Yours faithfully,

Details of mare sent to stud :

Name : ....

Sire : .... Dam : ....

Age : .... Height : ....

Number of previous foals : ....

Was she served last season? .... Date ....

Is mare to be sent : in foal, with foal at foot or barren? ....

Is the mare stitched? ....

Date last wormed? ....

List diseases (if any) against which the mare has been immunised? ....

To :

Miss A. Hardman,

Woodfield House Stud.

(Every precaution will be taken against accidents or disease, but no liability can be accepted therefore.)

3

SENDING YOUR MARE TO STUD

The stud season now officially starts on February 15th and ends on July 15th, but due to the fact that many ponies are foaled outside, their owners like them to foal in warm weather; the season for pony extends into late summer.

The gestation period for mares is 11 months or 340 days, although mares can foal a fortnight early and some a month late.

As previously stated mares come in season approximately every three weeks, therefore a mare served by a stallion but not holding to that service will usually come in season again three weeks later. These points must be taken into consideration when fixing a suitable date for sending a mare to stud.

If you intend to show your foal, as a foal or yearling, an early one i.e. born in February, March or at the latest early April is essential, as the bigger the foal the more likely it is to win a prize, always assuming its conformation is correct, even a few weeks makes a big difference to size during the first year.

Make sure your mare is healthy and in good condition before sending her to stud. Over-fat mares are notoriously difficult to get in foal, so if your mare tends to put on a lot of weight very easily, a course of slimming before sending her away will assist the stud enormously to get her in foal – it will also help if you inform them that your mare is inclined to put on weight. Thin mares in improving condition are usually easy to get in foal providing they are in good health and not thin due to worms or some obscure illness, once in foal they will often be seen to put on flesh rapidly – this is probably due to an improved ability in food assimilation which appears to take place once a mare is in foal.

A few days before you actually send your mare to stud worm her – this helps to prevent the stud's paddocks becoming horse

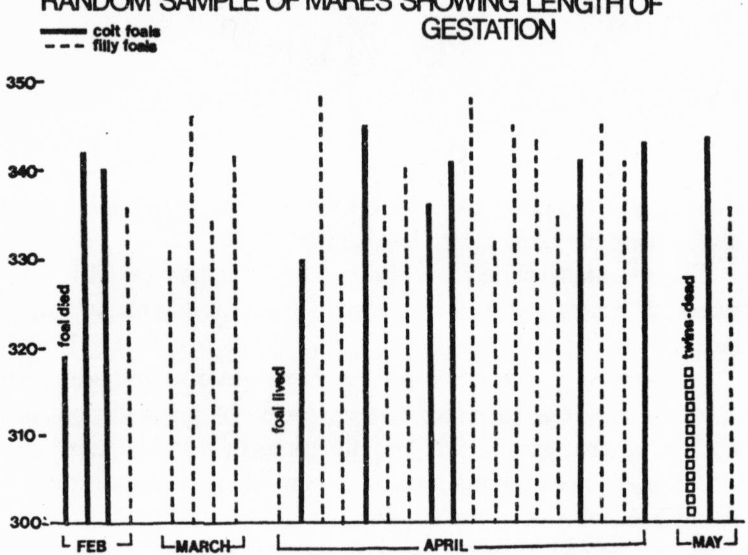

sick; also make sure you have had your mare's hind shoes removed as all studs make this a condition before they will allow the stallion to serve a mare and indeed before they turn her out with the other mares – as untold damage can be done by a shod horse.

Notify the stud as soon as possible of the exact day and time you intend to arrive with your mare. This is best done over the telephone, so that a time suitable to both parties can be arranged.

When taking your mare to the stud for any length of time

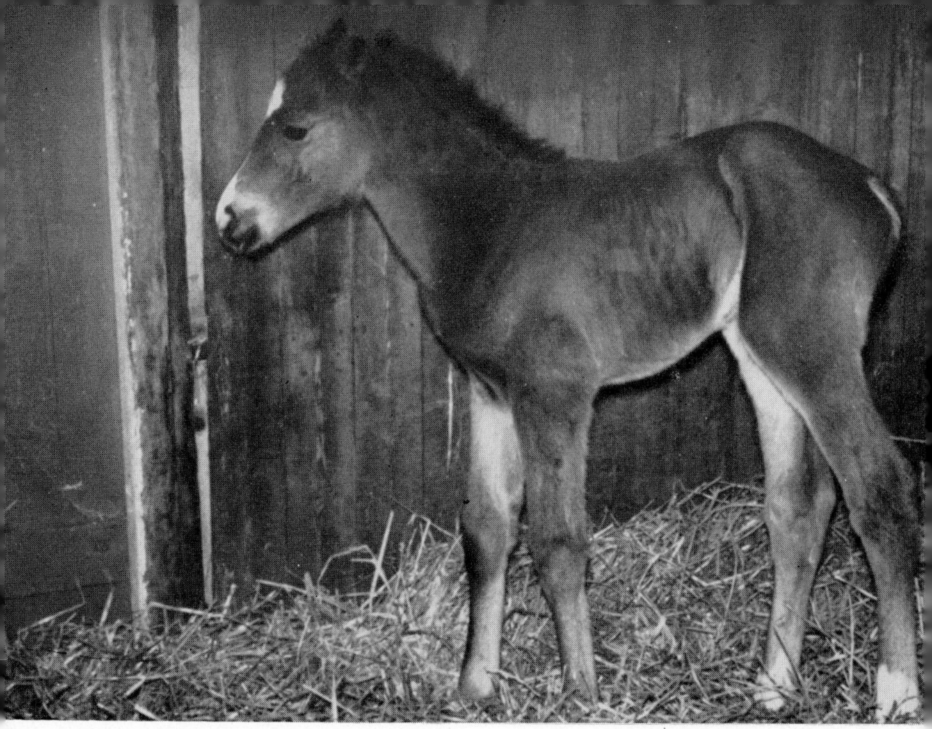

Above: A foal of your own . . . (*see* chapter 1). *Below:* A mare will sometimes accept an orphan foal and rear it with her own, but supplementary feeding of the foals is very necessary as the mare will not have enough milk for two. This is only to be recommended when a foster mother cannot be found (*see* chapter 1)

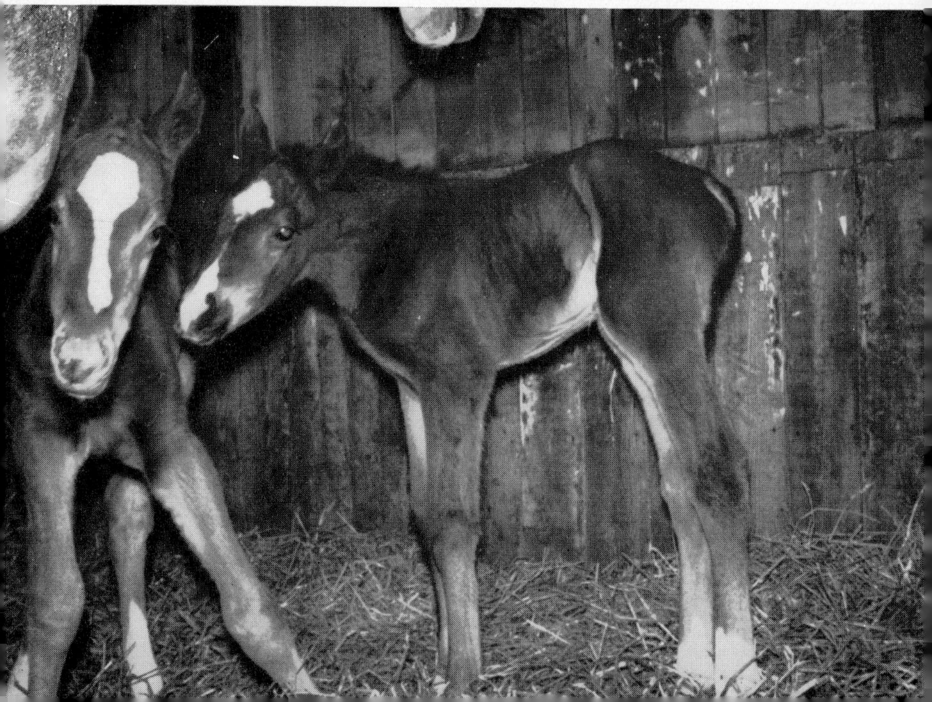

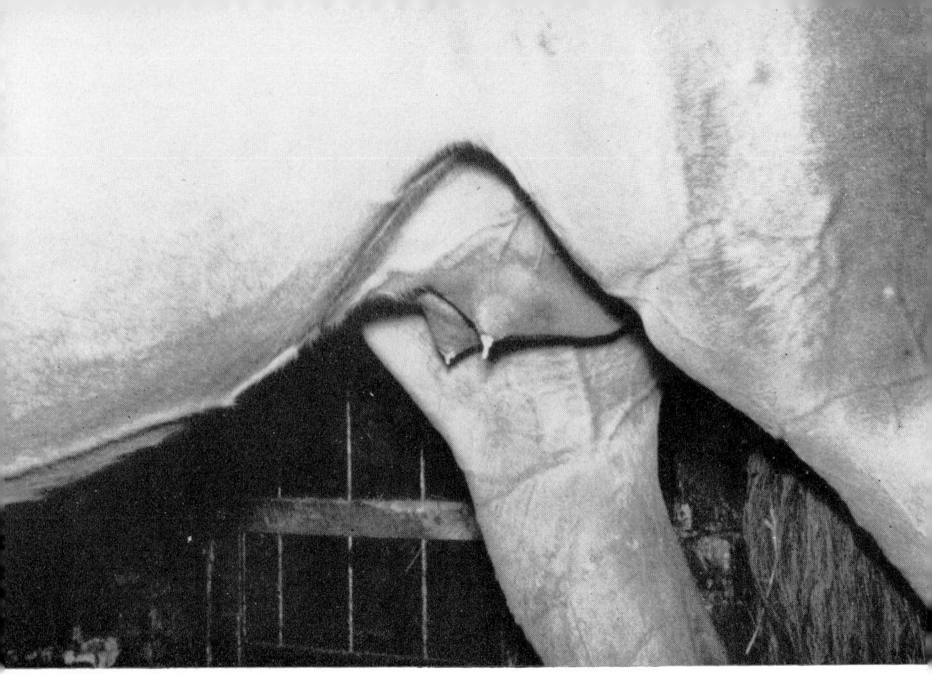

1. The mare's udder gradually gets larger during the last month of pregnancy and the muscles either side of her tail gradually get slacker. Candles of wax may appear on the ends of each teat. At this stage the mare could foal any moment. 2. In the case of some mares, the candles of wax disappear and they will run their milk from both teats before foaling

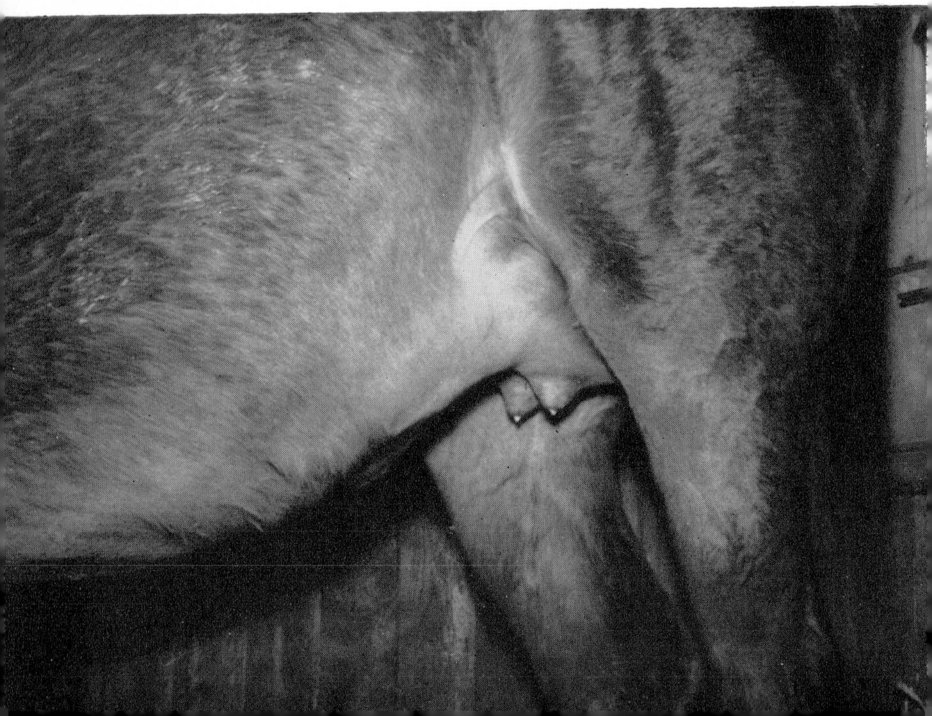

make sure she is wearing a leather headcollar which is exactly her correct size and preferably bearing a brass plate engraved with her name – in this way the headcollar will not be lost and the staff will be able to learn your mare's name easily.

Mares can only conceive if they are served while they are in-season, therefore a few weeks keep charges can be saved by taking her to the stallion as soon as she comes in use – it is better to take her the first time you see her on, as a long journey can sometimes upset a mare and put her off being in-season.

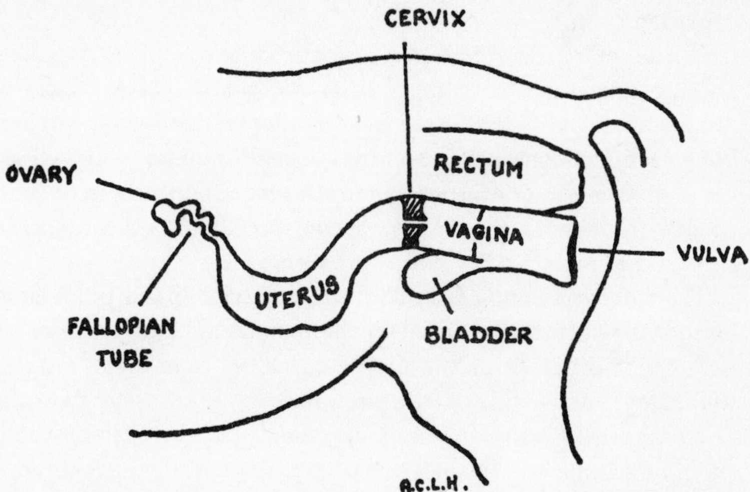

The visible signs that a mare is in-season are: She will call to other horses in the neighbourhood; if she is out at grass, on seeing another horse will often rush up to it and show very marked signs of interest. On coming into contact with other horses she may lean towards or against them, raise her tail and pass water, her vulva will be relaxed and will continually open and close; this is known as 'winking' if the lips of the vulva are parted the mucous membrane inside will be seen to be a moist red colour as opposed to the usual dry, salmon pink of a mare not in-season.

The mare has two ovaries, which are bean shaped and about the size of a small hen's egg, each is suspended by a ligament from the roof of the abdominal cavity about an armslength from the vulva.

Each ovary has several raised areas on its surface one being generally much larger than the others – these elevations are known as Graafian follicles which gradually ripen and in turn discharge a ripe egg (ovum), known as ovulation. The egg is discharged at the point on the ovary known as the ovulation fossa, which occurs in the middle of the concave side. As the egg ripens, so it becomes larger and softer, usually measuring from 2.5 cm. to sometimes as much at 7.0 cm. across before ovulation.

If the mare does not get in-foal, as one follicle ruptures another one starts to enlarge. After the rupture of a follicle a small round scar is left in the ovary – this gradually gets smaller but where fertilisation has occurred a small solid body known as the Yellow Body or Corpus Luteum forms around the scar and persists throughout pregnancy, during pregnancy other follicles rupture and more Yellow Bodies are formed.

The *Corpus Luteum* is in itself an endocrine gland producing hormones which are essential to the maintenance of pregnancy.

Follicles differ from the *Corpus luteum* in that they contain fluid. Eggs escape from a ruptured follicle about every 21 days and pass down the oviduct and fallopian tube where fertilisation should take place. If fertilisation does not occur, the egg remains in the fallopian tube and is gradually re-absorbed; only fertilised eggs enter the uterus. Ovulation generally occurs 24 hours before the mare stops being in-season. For successful fertilisation, live sperm should be present at the moment of ovulation as the egg dies 12-24 hours after being shed and as fertilisation must take place in the oviduct or fallopian tube, live sperm must therefore be present in the tube at this time – the fertilised egg then passes on down into the uterus and there pregnancy commences.

Sometimes when a mare is in-season the egg will ripen but instead of being liberated in the usual way it will remain on the ovary and the mare will stay in-season for some considerable time – sometimes for a month or more, in this state she cannot be got in-foal however many times she is covered until the egg is liberated. To facilitate ovulation the veterinary surgeon will inject the mare with Lutenising hormone which has the effect of liberating the egg and as long as there is live sperm present at the time of ovulation there is every chance that the mare will conceive. Lutenising hormone is a very safe hormone to use and has no carry-over effect to the next oestrus. This is common practise on most studs today, and usually gives very good results.

It is almost essential to leave your mare at the stud until she is no longer in-season. Most studs serve their mares every other day until they go off – this is due to the fact that although sperm from a very fertile stallion will live for as long as a week under ideal conditions inside a mare; under average conditions it probably only lives about three days and in old mares it sometimes only survives for 24 hours. As we have seen the egg is not liberated until the last 24 hours before the mare goes off and in order that it has the maximum chance of being fertilised active sperm must be present at the time of ovulation.

On arrival at the stud the stud groom will probably check your mare for any signs of a contagious disease, e.g. Strangles, coughing, lice, etc. If your mare is in season he may check to see if she is showing any signs of a vaginal discharge. As infected mares don't always show signs of discharge, most well-run studs require all the mares visiting their stallions to be swab tested before service, this prevents any chance of the stallion becoming infected and thus passing it on to the other clean mares.

The swab test is carried out very simply by the veterinary surgeon while the mare is in-season and the cervix is open, he inserts a speculum (a long hollow tube) into the mare's vagina, down which he passes a sterile swab on the end of a wire rod which is contained in a narrow bore metal tube, the swab is rotated gently within the cervix and withdrawn, it is then plated out in a laboratory and after incubation the resulting growth on the plates is examined for pathogenic organisms.

If a mare is infected a course of antibiotic irrigations while she is in season will usually put her right.

It is as well to note here, that any mare which is found to be barren on examination at the end of the year, is best swabbed on coming in-season, so that should she be found to be infected, she can be treated over the Winter period and thus much valuable time during the breeding season can be saved and an early foal obtained the following year.

Ideally leave your mare at stud for six weeks after her last service and have her examined for pregnancy before she returns home – in this way should she come in-season at an odd time outside her normal three week cycle she will be spotted by the stud groom who will be trying your mare with the stallion at least every other day after the first nine or ten days from service; whereas if she remained at home she could be missed.

When your mare has been covered by the stallion and is no longer in-season, you will probably receive a covering card from the stud which will notify you of the dates your mare visited the stallion. This is not to be confused with the actual covering certificate which is issued at the end of each season to mare owners, whose mares visited Thoroughbred or Arab stallions and are pure-bred or Anglo-Arab themselves. This latter certificate must be kept and sent in to the breed society concerned together with your annual returns when the time comes.

WOODFIELD HOUSE STUD

Dear

Your mare .... *was covered*

by .... *on* ....

Yours faithfully,

As far as annual returns are concerned, it is advisable to ask the stallion owner, for details.

Thoroughbred mares, registered in the General Stud Book must have a return made, on special forms supplied by Weatherby's for each year they are at stud. The return is made the year after covering, so that details of the foal can be given. THE RETURN MUST BE MADE EVEN IF THE MARE IS BARREN. Thoroughbred foals must be registered with Weatherby's by July 31st (in the year of birth) to qualify for the £5.50 fee for General Stud Book registration. After that date the fee goes up to £14.30, with a further increase to £19.80 on October 7th and to £30.80 on December 31st. Many other breed societies have similar schemes. Thoroughbred foals must have their markings recorded by a veterinary surgeon before they are 4 months old.

4

CARE OF THE MARE UP TO FOALING

It is a good idea to have a pregnancy test carried out on your mare before she leaves the stud, in this way you will know definitely if she is in foal and then if she proves barren and the season is not too far advanced further steps can be taken to get her in-foal.

If on the other hand your mare is already home and the stud season is over it is still a good idea to have her tested as in this way you will know how to deal with the mare: whether to hunt her or not, how to feed her and how much work to give her.

There are three tests available:

(i) Manual Test – carried out by a vet. at anytime from 40 days after the last service date, up to foaling.

(ii) Blood Test – a blood sample (30 cc.) is collected from the jugular vein by a vet. between 50-90 days after service and is sent away for testing. (Optimum time 70 days.)

(iii) Water Test – a sample of urine is collected by the owner anytime from 120 days after service and is sent away by a vet. for testing.

A method of collecting a urine sample for the water test:—

To obtain the best sample for testing this should be taken first thing in the morning after the mare has been left in a box over-night without water.

Equip yourself with a clean plastic bucket – a metal bucket is too noisy. Fill a second bucket with some warm water, rub plenty of soap on your hand and soap the inside of the vulva well – this sets up an irritation which causes the mare to urinate quickly. If possible take the mare into a different loose-box which has been occupied by another horse – allow the mare to smell the bedding, shake some straw under the mare and whistle, the moment the mare shows any signs of staleing put the bucket under her tail – collect enough to fill a medicine bottle, transfer the sample from the bucket to a suitable bottle, label this with your name, your mare's name, age, her last covering date, date of collection and your vet's name and address. Give the sample to your vet. It usually takes about one week for the results of both the blood and water tests to come back.

An alternative to a plastic bucket for the collection of a urine sample, is to fix an aluminium or plastic jug onto the end of a long stick. This apparatus can then be passed quietly behind the horse at the moment he stales, without frightening him unduly and without any chance of your being kicked.

(It should perhaps be noted here that when a urine sample is required for bacteriological examination it should be collected from mid-stream of the urination *direct* into a sterile container which should be corked immediately, with a sterile stopper.)

The accuracy of any of these tests lies solely with the person carrying them out. The main advantage of the Manual Test is that at the present time it can be carried out at a much earlier date than the other tests and that it is so far the only method of detecting the presence of twins or an under-sized foetus which may be re-absorbed probably by 9 weeks, though twins cannot be detected by manual examination later.

Actual union between the mare and foal is not complete until about 9 weeks after service; up to this time the foal is literally floating about inside the mare and can be lost at any time under unfavourable conditions but after 9 weeks the pregnancy becomes established and loss of the foetus is far more

CODE NO. OF MARE	POSITION OF RIPE EGG AT TIME OF OVULATION	POSITION OF FOETUS AT TIME OF MANUAL PREGNANCY DIAGNOSIS	
1	Left ovary	Right horn	*
2	Left ovary	Right horn	*
3	Right ovary	Left horn	*
4	Right ovary	Right horn	
5	Left ovary	Left horn	
6	Right ovary	Right horn	
7	Right ovary	Right horn	
8	Left ovary	Left horn	
9	Right ovary	Right horn	
10	Right ovary	Left horn	*
11	Left ovary	Right horn	*
12	Right ovary	Right horn	
13	Right ovary	Right horn	
14	Left ovary	Left horn	
15	Right ovary	Right horn	
16	Left ovary	Left horn	
17	Left ovary	Left horn	
18	Right ovary	Right horn	
19	Left ovary	Left horn	
20	Left ovary	Left horn	
21	Left ovary	Left horn	
22	Left ovary	Left horn	
23	Right ovary	Left horn	*
24	Left ovary	Right horn	*
25	Left ovary	Right horn	*

A random sample of in-foal mares showing the position of the foetus at 40 *days, related to the position of the ripe egg at service time. The alteration in position amounts to* 32 *per cent.*

unlikely. Consequently many mare owners have their mares re-checked after 9 weeks, especially in the early part of the season – when there is still time to get the mare covered again.

With regard to twins, as far as possible this can be prevented in the early stages by manual examination of a mare when she comes in-season and just simply not covering if two eggs are put up – but this does not of course prevent the possibility of one egg dividing and producing twins in this way.

It does, however, sometimes happen that although twins are produced they are not both carried through to parturition but that one at a very early stage fails to develop naturally and is re-absorbed while the other is carried to its full time. Thus when a vet. on rectal examination discovers twins he usually re-checks the mare a week or so later to determine if they are both developing normally; if this is the case and the owner so wishes it, the twins are washed out and a fresh start is made or they can be left – although twins are more often than not slipped before they reach their full time.

Thus should a mare be found to be carrying twins an owner has two courses of action open to him: either (i) leave them alone and hope that the mare may re-absorb one foal, or (ii) have the foals washed out by irrigation of the womb – in this case (if left to nature) the mare may not come in-season again for some months, probably due to the remnants of the foetus or of the cups which attach the developing foetus to the wall of the uterus, being retained and maintaining a high level of pregnant mare's hormone in the blood stream, thus preventing her from developing a follicle. However, an injection of the hormone Prostaglandin, given at the period in the mare's cycle when she has an 'active' yellow body (*Corpus luteum*), will cause her to come in-season again within a few days.

The main disadvantages of twins are: (i) The great possibility that they will not be carried to full time and thus slipped or born dead: (ii) If they are born alive they tend to be small and remain stunted all their lives. Although good feeding can help to make good the pre-natal low level of nutrition, it can never quite make up for the early deficiency: (iii) The danger of damage to the mare during foaling and even possible loss of the mare and the risk of not being able to get her in-foal again until the following year: (iv) The possibility that the mare will not be able to rear two foals and the consequent bother of obtaining a foster mother for one of them or alternatively rearing it by hand.

About half way through the gestation period the foal can often be seen to kick inside the mare, this is particularly noticeable if the mare is left without water for a few hours and then given a bucket of cold water to drink.

Other good signs that a mare may be in-foal are that the mare will often appear more docile and will begin to drop noticeably on her underline. This can best be seen if photographs are taken of the mare at approximately monthly intervals and compared one with another.

When you collect your mare from the stud it is probably better not to load her into a very narrow compartment where her sides are touching the sides of the trailer or horse-box, just as it is better not to strap an in-foal mare vigorously, both these things can cause a mare to strain and thus run the risk of losing her foal; light grooming to remove mud, etc., is of course all right.

If the mare returns home soon after service do not ride her hard or in any way upset her too much; when she is safely in-foal you can carry on as usual but the mare herself will gradually get slower as she gets more heavy, until towards the end you will probably find it easier to turn her out in the field than to ride her, although quiet exercise is essential for an in-foal mare right up to foaling.

If foaling mares cannot be turned out each day into a paddock for exercise, or if on being turned out they tend to stand around and not exercise themselves, they should, whenever possible, be led out in hand for about half-an-hour each day for the last month or two before foaling, when they are too heavily in-foal to ride.

Exercise is of great importance where the brood mare is

concerned. Lack of exercise just before foaling, can lead to difficult foalings and retained meconium in the foal. Therefore, even in bad weather mares should have some exercise each day. The only exception to this rule would be when the ground is covered with ice or snow, in this case the risk of a fall and probable abortion, would far out-weigh the lack of exercise, so the mare should be left in her loose-box.

If you have the facilities, a straw ring can be made for winter exercise. To do this, instead of taking all the wet straw and droppings to the 'muck' heap each day they can be spread thickly in a ring, round which the horses can be walked. Straw is not very safe when placed on ice as it tends to slip unless is is very thick. A good method of rendering concrete paths safe to walk a horse on in icy weather is to scatter plenty of ashes or sand on them; this must, however, be renewed each day as it sinks into the ice as the latter melts and the surface becomes slippery again.

Mares that are believed to be in-foal should not be turned out with or be able to touch noses with geldings or entire colts until they are at least half way through their gestation period and better still, never. The best companion for an in-foal mare is another in the same condition; failing this an older barren mare, rather than a young mare or filly which might gallop about, though these would be preferable to a colt or gelding which would probably tease the mare, cause her to slip foal and bring her in-season again.

Brood mares should be wormed regularly and some time within the last two months before foaling, as some foals will play with and lick or eat their mother's droppings, so this will reduce the risk of the foal picking up a worm infestation from its dam. Thiabendazole is at the present time the best and safest drug to use. Other drugs are not safe to use near foaling time.

Once mares are brought in for the winter, they should be supplied with a salt lick and ad lib hay and water. Mares that are in-foal will need more bedding than usual and must be kept bedded down all the time as they will often lie down during the day for a rest. Mares in foal do of course require more food than barren mares. They should be fed three or four times a day, i.e. a medium dry feed in the morning, a small feed at mid-day and a large mash or dry feed at night.

All breeding stock should be vaccinated against flu and Tetanus; brood mares should receive an annual booster injection preferably about four weeks before foaling in the case of Tetanus. In this way, the foal will receive protection through its mother's milk for the first two months, after which it can be vaccinated itself.

A few months before the mare is due to foal you must make up your mind what to do with her. There are three possibilities:

(i) Send her to the stud to foal down.
(ii) Foal her yourself in a loose-box.
(iii) Let her foal outside in a field.

Method (i) is probably the best, if you have had no previous experience of foaling and intend to breed from the mare again that year – most studs are equipped with special foaling boxes and sitting-up rooms with experienced staff in attendance all the time: both day and night.

Method (ii) calls for a suitably large loose-box (16 x 16 feet) for a Thoroughbred mare, with an electric light and heater and a suitable observation point for the watcher – preferably somewhere where the mare cannot see or hear you.

Method (iii) is only to be recommended where alternative loose-box accommodation is inadequate or cannot be kept really clean. Foaling outside is certainly not to be recommended early in the year when the weather can turn cold and wet suddenly, as new-born foals do not have a protective layer of grease in their coats for about the first fortnight and could therefore get pneumonia. A further drawback to this method is that most mares foal between 11.0 p.m. and 3.0 a.m. when everything is quiet and they imagine that they are alone, so should they get into any trouble while foaling, it would be very difficult to give help to the mare in the pitch dark even if you happened to see her.

If the mare is not already insured on an annual basis she should be covered for foaling and 30 days afterwards; if she is insured annually, the premium should be increased to cover the act of foaling. Once the mare has been tested in-foal the unborn foal which she is carrying can be insured against abortion, being born dead, or dying within 48 hours after birth, for a premium of 17% or more according to the mare's record.

Thoroughbred mares can be insured through Weatherby's, Sanders Road, Wellingborough, Northamptonshire NN8 4BX, simply by sending them the name of the mare and her foaling date. Other mares, including Thoroughbreds, can usually be insured through the various Bloodstock Agencies at favourable rates.

5

FOALING

Choose the largest and warmest loose-box to foal your mare in; the floor should be concrete – roughened to prevent slipping, so that it can be kept clean and disinfected easily. All holes or gaps in the lining boards which might allow a foal's hoof to slip through and possibly get trapped should be made up. All draughts must be eliminated by placing rolled sacks across the bottom and down the sides of the door. To help to protect the foal from injury during its first struggles to get to its feet, the box should be well bedded down with good clean wheat straw and the straw built up round the sides of the box to a height of two feet including the area round the door. The box must be mucked out every day and ideally, the bed left up while the mare is out at grass, to allow the floor to dry out. Cleanliness at the time of foaling is very important as dirty conditions can cause Joint Ill in the foal. (See Chap. 11, p. 117.)

A loose-box which is sited so that its interior can be seen from your house is ideal, in this way you can leave the electric light on in the box all night and watch the mare through a suitable window without disturbing her and without her knowing she is being watched. The top door of the loose-box should be left open to give a better view of the mare. Alternatively a box should be chosen with an adjoining saddle room or feed house which can be used as a sitting-up room. A trap door should ideally be built into the wall to enable the person who

DIAGRAM OF FOETUS AND ITS MEMBRANES IN THE MARE'S UTERUS.

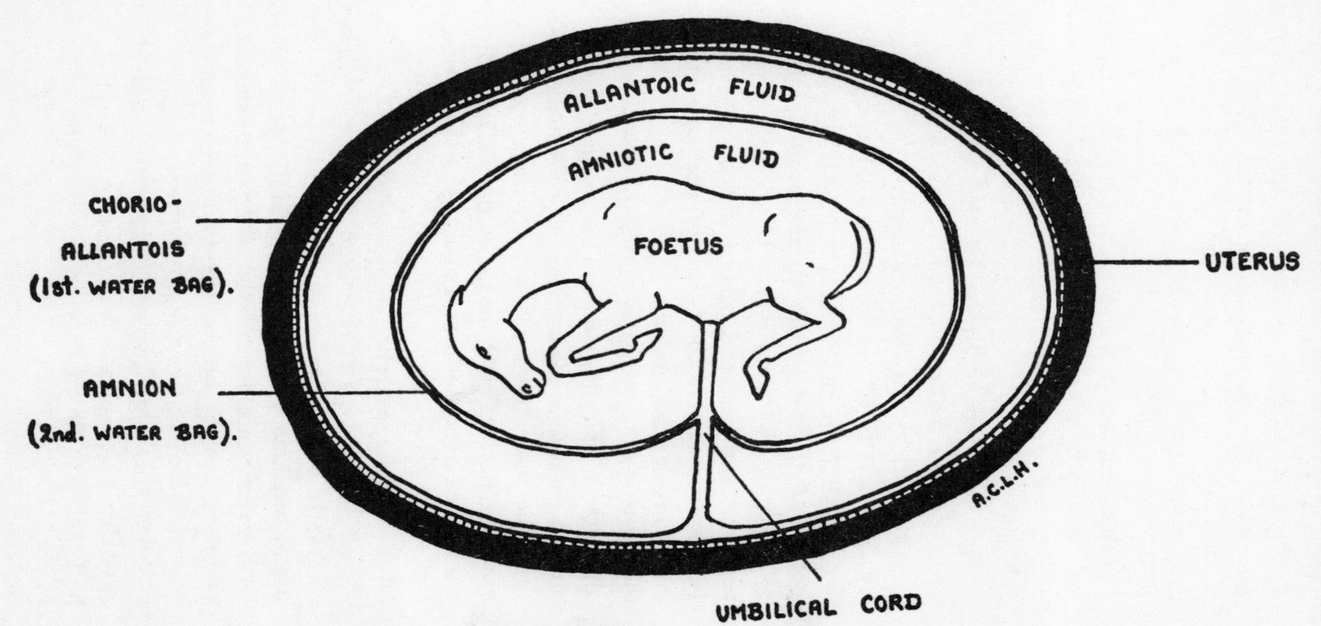

is sitting up to keep a watch on the mare without needing to go into her at regular intervals. All large commercial studs have special sitting-up rooms adjoining their foaling boxes and many of these are equipped with closed circuit T.V.

Other important points are: The water bucket in a foaling box should be plastic rather than metal – a metal bucket standing on the floor could cause damage if the foal happened to fall against it.

Hay nets should always be tied up high to eliminate any chance of the foal putting a leg through the net or the hay can, if preferred, be left loose in a corner of the box. When tying up hay nets it should be noted that the net can be tied up far higher if the draw string is passed via the ring through the bottom of the net to the highest point up the side of the net, at which it can comfortably be secured: this is best done by making a slip knot and passing the end of the rope through the loop, so that should the horse pull the loose end of the rope, the knot will not come undone.

As the mare gets nearer to foaling she will tend to get slower and slower and less inclined to exercise herself when turned out. As exercise is essential for an easy foaling it is advisable to walk the mare out for about 30 mins. each day if you have time. As parturition approaches her udder will be seen to get gradually larger but will usually go down again after exercise but for the last ten days or so before foaling it will remain large even after exercise.

Just before foaling the mare will tend to get very restless, constantly wandering round her box, possibly getting up and down, swishing her tail and kicking up at her belly. Minor birth pains can sometimes be detected as long as a fortnight or so before foaling but as the time gets nearer they gradually become more frequent and pronounced.

When the mare seems to be getting near to foaling it is advisable to telephone your veterinary surgeon to ascertain if he is going to be on duty, in this way you won't waste time, if

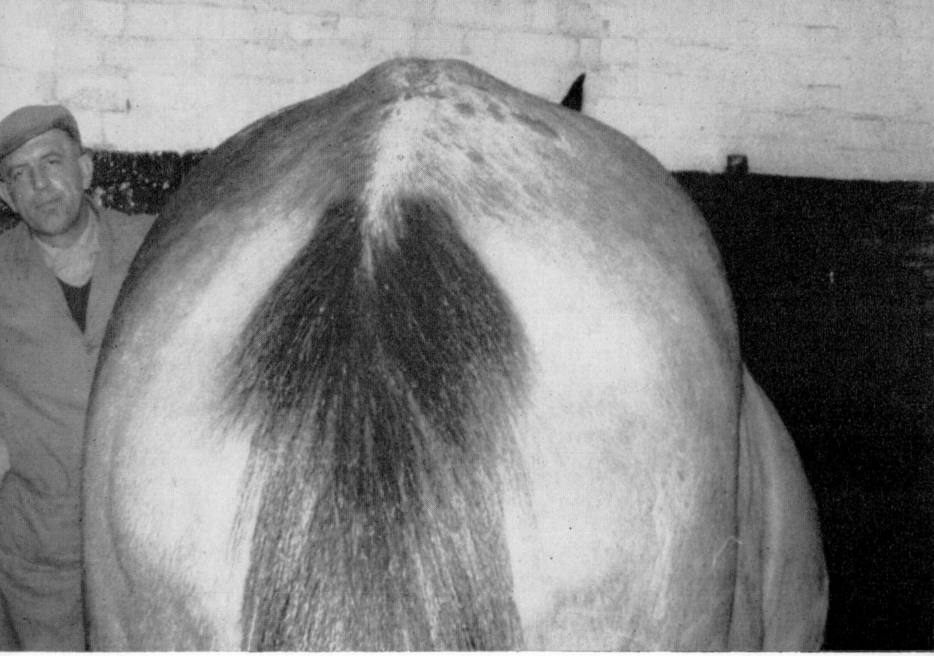

Above: As parturition approaches so the muscles either side of the tail gradually sink and become slack. At the same time the vulva will gradually lengthen and swell. When the mare actually starts to foal, she will often walk her box and paw at the straw. She may start to sweat profusely (*see* chapter 5). *Below:* She may get down to foal or may commence foaling in the standing position, only going down as the foal begins to arrive. The first sign of foaling is the breaking of the water bag (placenta), shown by a flow of brown water (allantoic fluid) from the vulva. This is soon followed by the appearance of the amnion containing some amniotic fluid (*see* chapter 5)

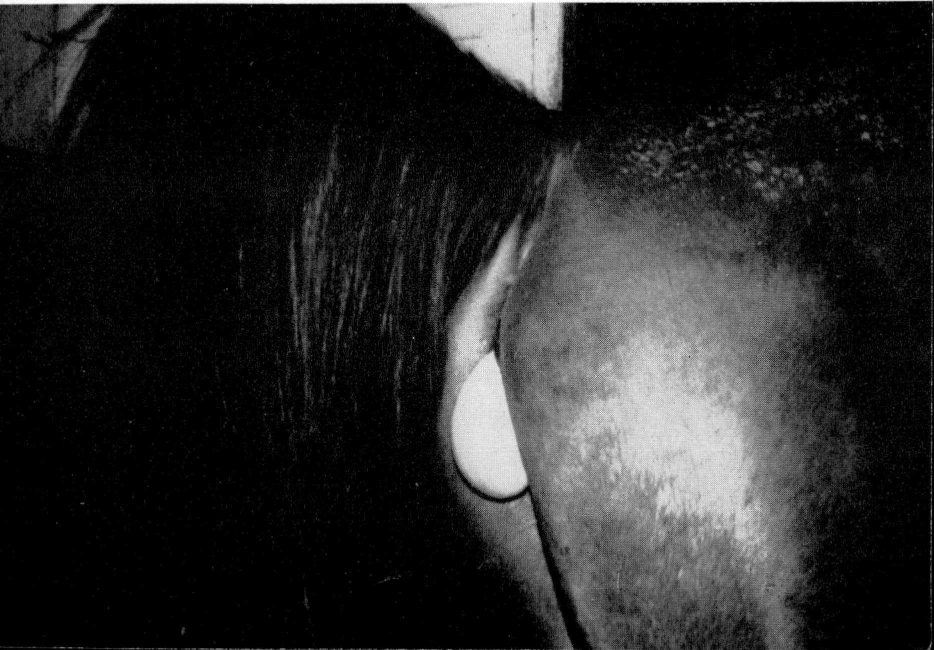

Above: Soon afterwards the front feet will be seen inside the amnion. These should be checked to make sure that one foot is coming in advance of the other – this allows the foal's shoulders to come through the pelvic arch easily. Should the attendant need to pull on the foal's legs, one must always be kept in advance of the other for this reason (*see* diagram on page 48) (*see* chapter 5). *Below:* At this stage the mare may get to her feet only to go down again on her other side. This is thought to be a natural method of correcting a faulty position in the foal. The foal's head will begin to appear between the front legs – in a diving position (*see* chapter 5)

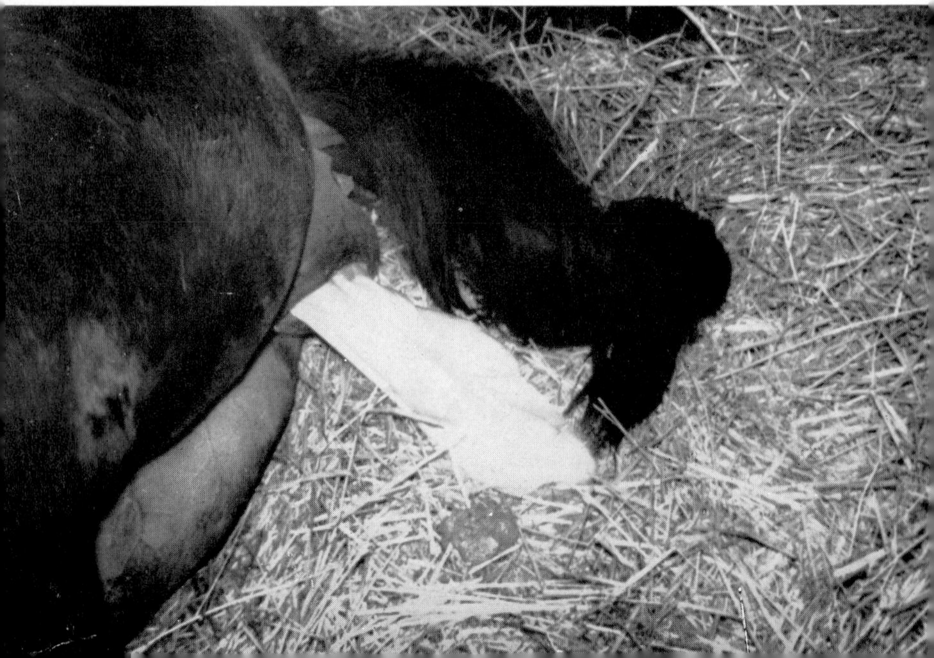

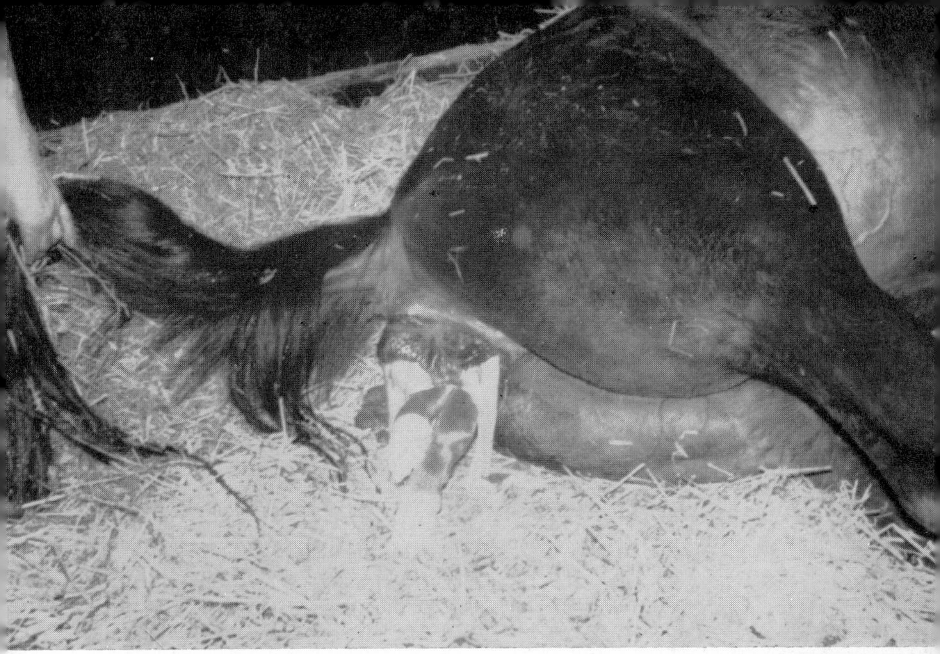

Above: The amnion will often break automatically as the front feet appear, but if it does not bear naturally it should be broken by the attendant as soon as the head is out. If the amnion remains un-broken when the cord severs the foal will suffocate. The foal itself will often break the amnion by arching its neck and striking with its front feet. The foal's hooves are covered on their edges by a softer pad which protects the mare from injury. These fall off as soon as the foal gets up (*see* chapter 5). *Below:* Once the shoulders are out the mare will often rest for some time before completing the delivery (*see* chapter 5)

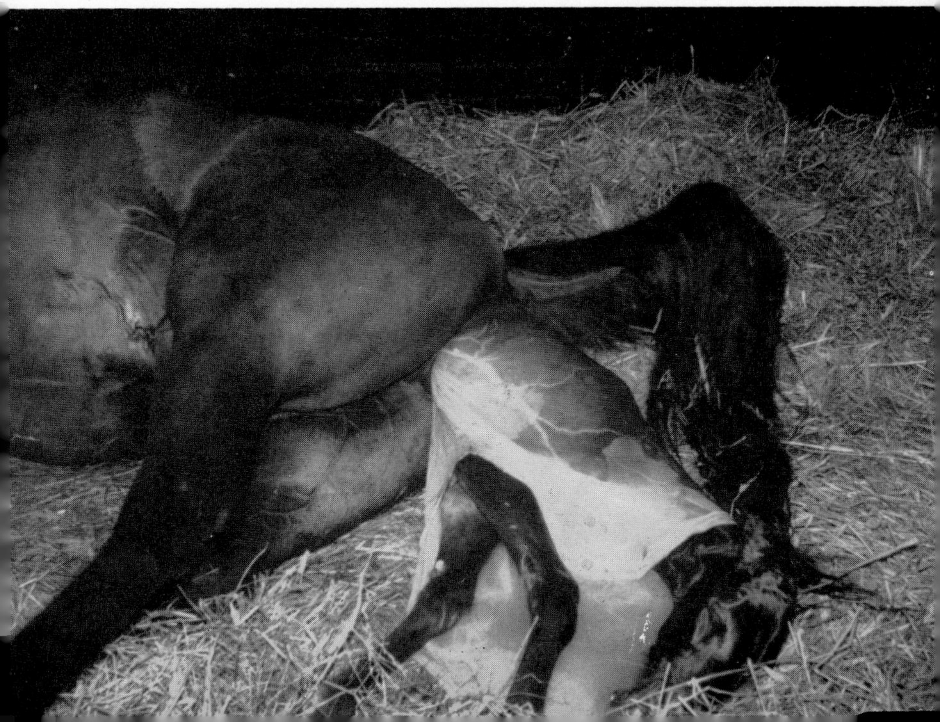

Once delivery is complete the mare and foal should be left to rest. On no account should the cord be broken artificially, as a considerable amount of blood flows through the cord into the foal at this stage (*see* chapter 5)

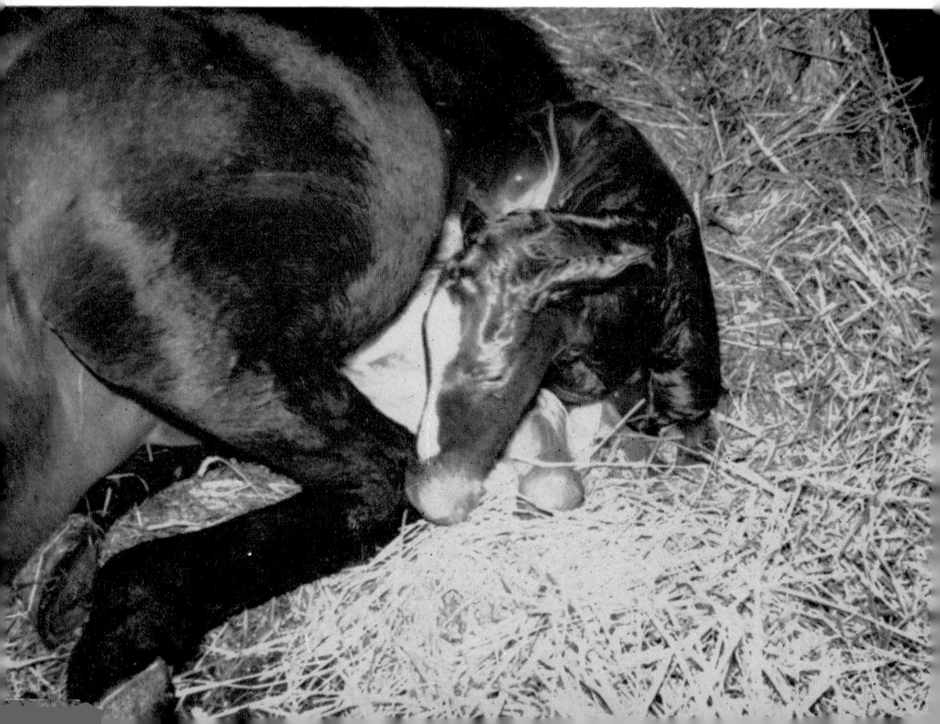

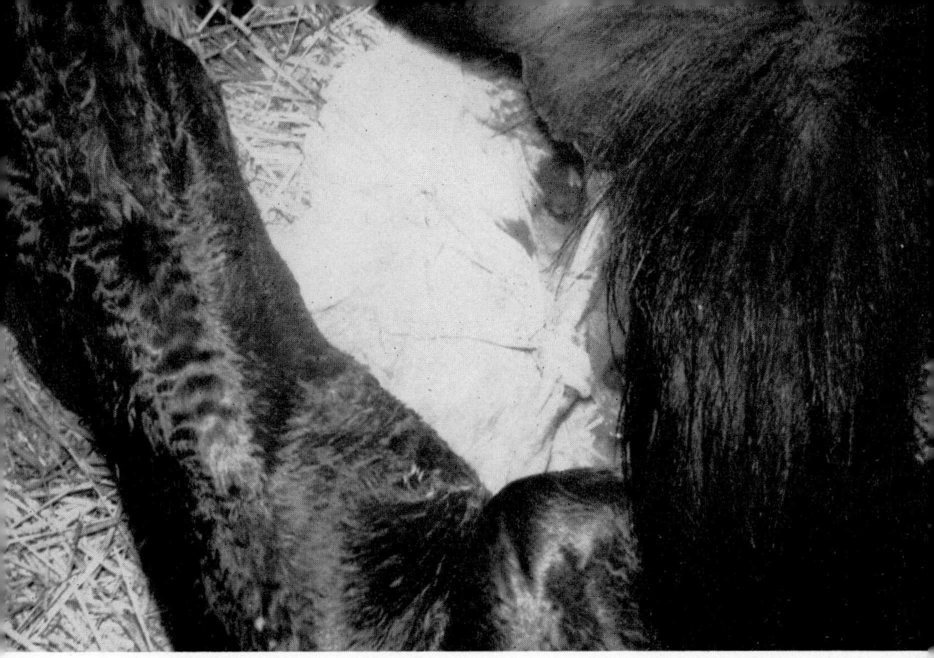

Above: The foal receives its entire oxygen supply from the blood flowing through the cord from its mother. When the cord breaks the foal's breathing must be fully established as it then has to breathe entirely on its own. The umbilical cord constricts as soon as it breaks naturally, thus preventing any loss of blood, but if it is broken or cut artificially considerable loss of blood from the foal can result (*see* chapter 5). *Below:* It is a very good idea to pull the hind legs clear of the mare's vulva without breaking the cord, as this prevents any damage to the mare when the foal begins to struggle to its feet – although some authorities maintain that the presence of the feet help to keep the mare down (*see* chapter 5)

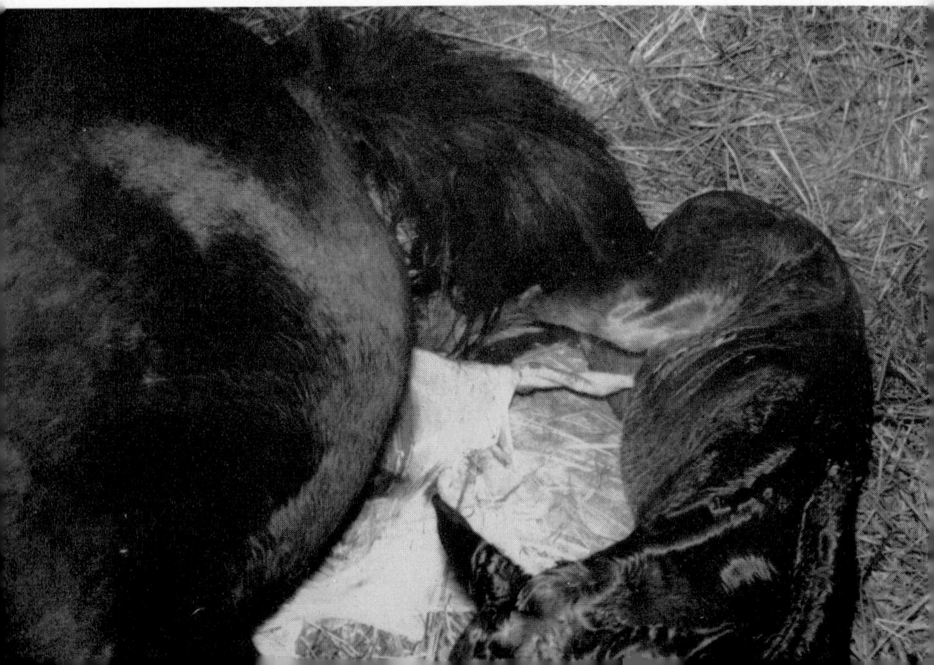

Gradually the foal will struggle to get up and the cord will break naturally. Once it has broken, a packet of Sulphonimide should be shaken over the stump. This should be repeated again the next day to prevent infection, such as Joint Ill (*see* chapter 5)

Above: The foal should be dragged round to the mare's head so that she can lick it dry. The mare will probably remain down for at least half-hour after foaling. *Below:* The after-birth should be tied or knotted up to prevent it flapping round the mare's hocks (*see* chapter 5)

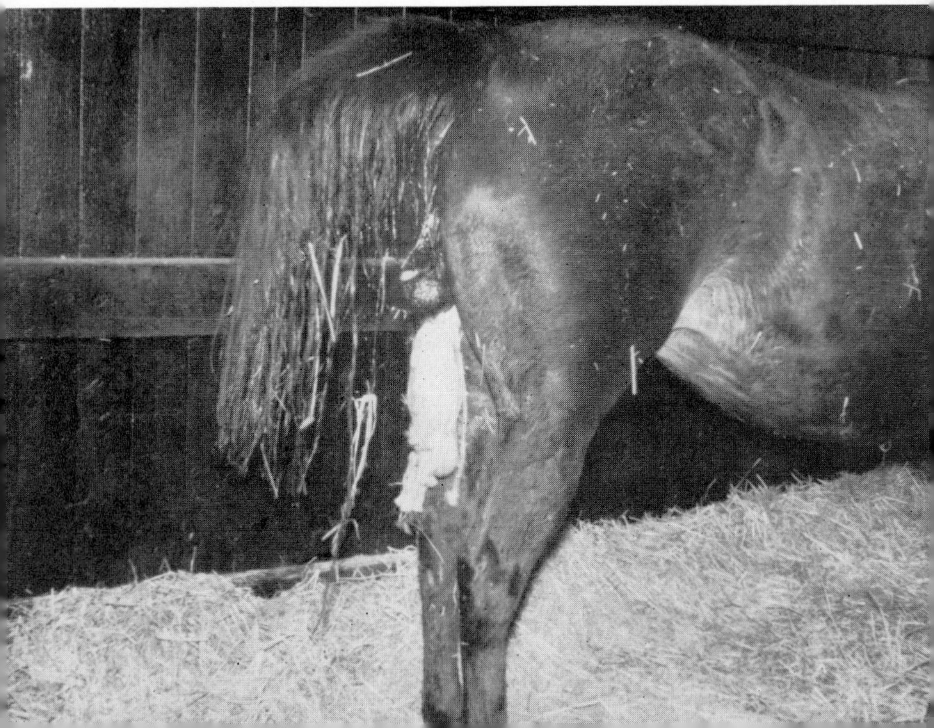

The after-birth or Placenta. (1) Main body of the chorio allantois (chorion) membranes (2) Pregnant horn (3) Non-pregnant horn (4) The amnion. When checking to see that all the after-birth has been expelled, lay the membranes out as shown above, and especially note that *both* horns are intact (*see* chapter 5)

Once the mare and foal get to their feet, the foal will immediately look for the mare's udder and should manage to find the teats within an hour (*see* chapter 5)

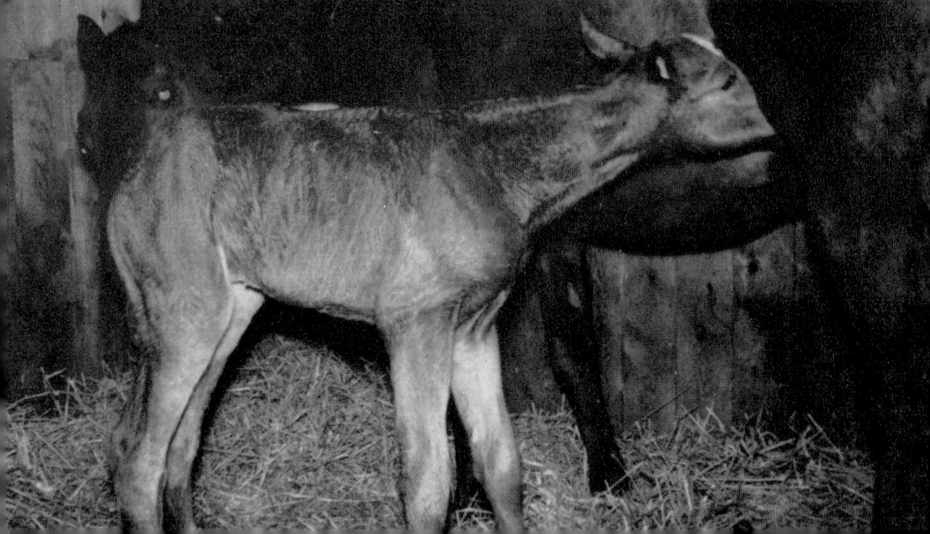

there is an emergency, trying to find a veterinary surgeon who is at home.

The mare's udder gradually gets larger during the last month of pregnancy until the final week when it will become shiny and remain large even after exercise. Candles of wax may appear on the ends of each teat; at this stage the mare could foal any moment.

In the case of some mares, the candles of wax disappear and

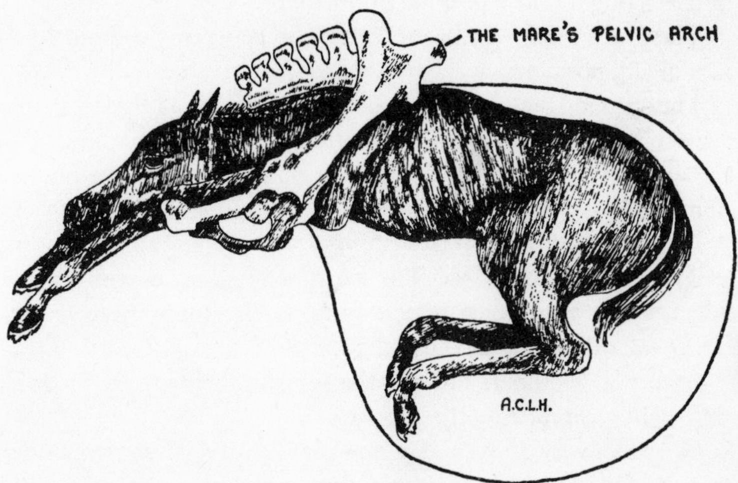

Diagram showing a correct presentation of the foal

they will run their milk from both teats before foaling.

As parturition approaches so the muscles either side of the tail will gradually sink and become slack. At the same time the vulva will gradually lengthen and swell.

When the mare actually starts to foal, she will often walk her box and paw at the straw. She may sweat profusely.

She may get down to foal or may commence foaling in the standing position, only going down as the foal begins to arrive. The first sign of foaling is the breaking of the water bag (placenta), shown by a flow of brown water (allantoic fluid) from the vulva. This is soon followed by the appearance of the amnion containing some amniotic fluid.

Soon afterwards the front feet will be seen inside the amnion. These should be checked to make sure that one foot is coming in advance of the other – this allows the foal's shoulders to come through the pelvic arch easily. Should the attendant need to pull on the foal's legs, one must always be kept in advance of the other, for this reason. (See diagram on p. 49.)

At this stage the mare may get to her feet only to go down again on her other side. This is thought to be a natural method of correcting a faulty position in the foal.

The foal's head will begin to appear between the front legs – in a diving position.

The amnion will often break automatically as the front feet appear but if it doesn't break naturally it should be broken by the attendant as soon as the head is out. If the amnion remains unbroken when the cord severs the foal will suffocate. The foal itself will often break the amnion by arching its neck and striking with its front feet. The foal's hooves are covered on their edges by a softer pad which protects the mare from injury. These fall off as soon as the foal gets up.

Once the shoulders are out the mare will often rest for some time before completing the delivery.

Once delivery is complete the mare and foal should be left to rest. On no account should the cord be broken artificially, as a considerable amount of blood flows through the cord into the foal at this stage.

The foal receives its entire oxygen supply from the blood flowing through the cord from its mother. When the cord breaks the foal's breathing must be fully established as it then has to breathe entirely on its own. The umbilical cord constricts as soon as it breaks naturally, thus preventing any loss of blood, but if it is broken or cut artificially considerable loss of blood to the foal can result.

It is a very good idea to pull the hind legs clear of the mare's vulva without breaking the cord, as this prevents any damage to the mare when the foal begins to struggle to its feet – although some authorities maintain that the presence of the feet help to keep the mare down.

Gradually the foal will struggle to get up and the cord will break naturally. Once it has broken, a packet of sulphonamide should be shaken over the stump. This should be repeated again the next day to prevent infection, such as Joint Ill.

The foal should be dragged round to the mare's head so that she can lick it dry. The mare will probably remain down for at least half-an-hour after foaling.

The after-birth should be tied or knotted up to prevent it flapping round the mare's hocks.

The after-birth (*placenta*) gradually separates from the uterus and is expelled by the mare usually within an hour of foaling. If the after-birth has not been dropped by the morning the veterinary surgeon should be called immediately to remove it, otherwise septicaemia may set in. Once expelled the after-birth should be examined carefully to make sure that all the parts are present and no piece has remained inside the mare which could cause septicaemia, after examination it should be removed from the loose-box and buried.

Once the mare and foal get to their feet, the foal will immediately look for the mare's udder and should manage to find the teats within an hour. The attendant should not leave the foal until he has made sure that it has had a drink. Weaker foals must be supported to enable them to drink.

Occasionally some maiden mares are very ticklish in the region of the udder and need to be twitched with a front leg held up before they will allow the foal to suck; they are usually all right once the foal has sucked and the udder pressure has been released a little.

It is very important that the foal should receive the mare's colostrum – or first milk, as the foal, unlike the human baby, does not receive any immunity to disease before it is born but receives its resistance to infection entirely from this fore-milk but it can only absorb the antibodies for about 48 hours after birth, after which time it starts to manufacture its own antibodies.

When your mare foals – if there is any variation from the normal foaling procedure described, you should notify your veterinary surgeon immediately as if there is anything wrong, speed in getting veterinary assistance is imperative.

Next morning the foaling box must be mucked out as it will be very dirty and wet. This should be done with the door shut and in sections. Each area mucked out should be bedded down again immediately to prevent any chance of the foal slipping up on the wet floor.

If the weather is warm and sunny there is absolutely no reason why the mare and foal should not go out into the paddock the day after foaling. If the mare has not been out in the paddock for several days she should be led around for about 15 minutes before she is allowed to go free, as many mares will kick up their heels on first being let loose and several foals have been injured in this way.

As from the first day the foal should be handled and taught to lead. Within the first few days the foal should be pushed gently round the box; to do this one hand must be placed round the foal's chest and your other hand round its tail; in this way the foal can be led anywhere. Once it has got the idea of moving around under your guidance a small soft head-collar should be put on the foal and it can then be led out with its mother in front, if the weather is fine, a hand being kept round its hind-quarters all the time to prevent the foal from running backwards and throwing itself down.

During this period the foal's feet should be picked up. The sooner this is done the less trouble you will have with it later. Make sure the foal's head-collar fits it well and has plenty of adjustment to allow for growth.

The mare comes in-season again any time from seven days onwards after foaling. Should you wish to get her in foal again the next year, it is as well to send her away to the stallion about 5 or 6 days after foaling, to give her time to settle down before service.

Random sample of 27 foaling mares.

CODE NO. OF MARE	DAYS ON WHICH MARE WAS SERVED AFTER FOALING	IN-FOAL OR RETURNED	RESULT OF SUBSEQUENT SERVICE
1	9 days	In-foal	
2	7, 8 days	RETURNED	In-foal
3	10 days	RETURNED	In-foal
4	8, 9, 10 days	RETURNED	In-foal
5	10, 12 days	In-foal	
6	9 days	RETURNED	In-foal
7	10 days	In-foal	
8	10 days	RETURNED	BARREN
9	9 days	RETURNED	In-foal
10	10, 11 days	In-foal	
11	10, 12 days	RETURNED	BARREN
12	11, 14 days	RETURNED	BARREN
13	8 days	RETURNED	In-foal
14	9, 11 days	In-foal	
15	9 days	In-foal	
16	8, 10 days	RETURNED	BARREN
17	10 days	RETURNED	In-foal
18	9, 11 days	In-foal	
19	9, 10 days	RETURNED	In-foal
20	9, 11 days	In-foal	
21	8 days	RETURNED	In-foal
22	10 days	In-foal	
23	8 days	RETURNED	In-foal
24	8, 9 days	In-foal	
25	9 days	RETURNED	BARREN
26	9 days	RETURNED	In-foal
27	11 days	In-foal	
		40.7% In-foal	68.75% In-foal

It is, however, not absolutely essential that a mare should return to the stallion for her foaling heat, as, contrary to common belief, this is the time of lowest fertility due to the mare in some cases not having cleaned up properly after foaling – statistics have shown that fertility at this time is as low as 40% of the total number of mares covered as compared with about 67% for later services.

Occasionally some mares only exhibit external signs of oestrus during their foaling heat and do not show signs of being in-season again, as long as they are suckling their foals. In these cases the mare should be covered at the foaling heat and examined for pregnancy as early as possible. Should the mare prove to be barren, she should be examined by a veterinary surgeon at least twice a week until she shows internal signs of oestrus, as soon as there is a ripe egg present, the mare should be covered. Some other mares do not show any signs of oestrus even at their foaling heat – in this case the mare should be examined by a veterinary surgeon from seven days onwards after foaling and covered by the stallion the moment a ripe egg is present. Very good results have been obtained from services under these conditions.

Mares should only be covered at the foaling heat if they are absolutely clean and are not bruised internally.

When the mare comes in-season for the first time after foaling, you will probably notice that the foal's dung has become loose. This is a perfectly natural occurrence and absolutely nothing to worry about. In order to prevent the foal's coat from dropping out in the region of the dock, any wet dung should be washed off and vaseline smeared over the entire area in contact with the dung.

If the mare has not been injected with Tetanus Toxoid about three weeks before foaling the foal should be injected with Tetanus Serum as soon after birth as possible and again a month later. This will probably happen while the foal is away at stud and will give it protection until it is old enough to be injected with the Toxoid. In case the stud do not inject their foals as a routine practice, it is as well to notify them that you wish to have your foal done – this is particularly important in some areas of the British Isles where Tetanus is very prevalent.

For the first few days after birth the foal should be checked to make sure it is passing its droppings easily. At the time of birth the foal has a hard black substance in its rectum, known as meconium – these are the faeces which have accumulated during gestation and which must be passed before normal functioning of the bowel can commence. The mare's colostrum is a natural purgative but should the foal be seen to strain or have any symptoms of colic a tablespoonful of liquid paraffin or castor oil should be administered without delay but should the foal appear to be in pain it is probably the wiser plan to summon a veterinary surgeon immediately rather than let the foal get any worse. (See chapter 11, p. 119.)

The act of foaling is generally considered to be made up of three distinct stages: the first stage occurs when the mare begins to feel pain, she starts to walk round, paws the ground and may sweat up. This stage can take up to four hours. The second stage starts with the breaking of the water bag and the flow of allantoic fluid and extends until the umbilical cord breaks. Under natural conditions this usually takes from 7–42 minutes.

The allantoic fluid can be determined from urine by the fact that it is usually darker in colour. At first only a small quantity escapes and this is not accompanied by the usual 'winking' which normally follows urination but is followed instead almost immediately by a larger quantity of fluid which gushes out of the vulva.

The third and last stage commences with the breaking of the umbilical cord and finishes with the complete expulsion of the after-birth. This can take from 0 minutes to 10–12 hours, after which time a veterinary surgeon should be called in to remove it manually, preferably first thing in the morning and before 12 hours are up. The second stage should not extend beyond about 40 minutes without veterinary aid being sought, to make quite sure that nothing is wrong. Often the first and second stages together only take about 90 minutes. If you suspect that there is something wrong and telephone your veterinary surgeon, he may advise you to get the mare up onto her feet and walk her round the box until he comes – to prevent her from straining any more and so exhausting herself.

6

FEEDING OF BROOD-MARES AND YOUNGSTOCK

The ultimate objective when breeding horses is to produce the very best possible foal from the stallion and mare available.

The two most important periods in any young horse's life are, in order of priority:

(i) The eleven months spent inside the mare (in utero).

(ii) His first winter – which is probably after weaning.

The period spent 'in utero' is by far the most important as this is the time which determines the size, make and shape of the foal at birth. Starvation of the foetus for any reason cannot entirely be made good after birth. Therefore, in order to obtain the best possible foal from your mare, she must be fed well throughout the whole period of gestation. Brood mares which are due to foal in the early part of the year, should be given a corn feed at grass from about the beginning of August onwards, i.e. the moment the feeding value of the grass begins to decline.

This is best done from wooden field troughs, which stand on legs and cannot be tipped over by horses pawing at them while feeding. If more than one horse is to be fed in the same paddock, each horse must be given his own trough and these should be placed far enough away from one another to prevent kicking.

When several animals are turned out together in one field one horse will become the boss and the others will assume dominance over each other in descending order. Until social

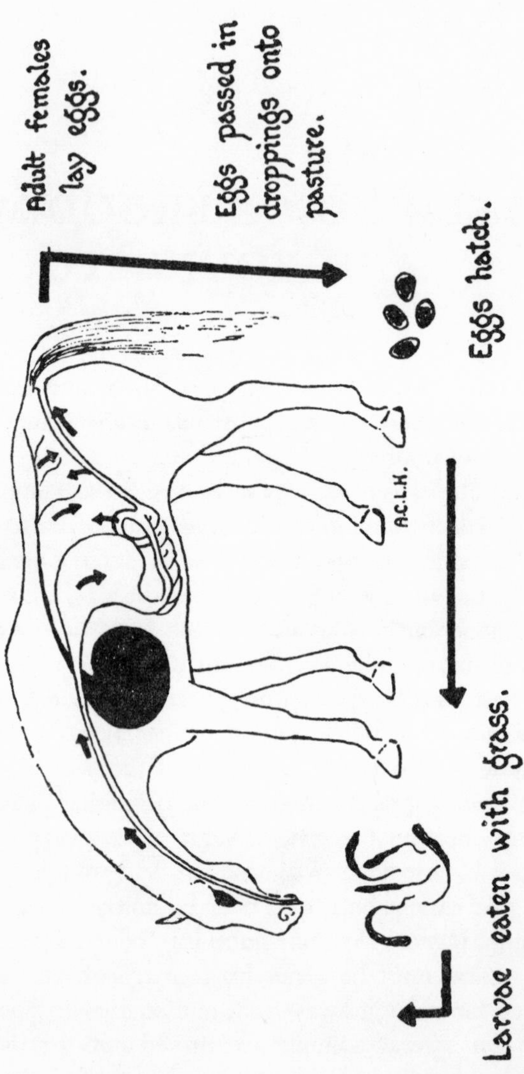
Larvae migrate through gut wall, into blood vessels
and return to large intestine to become adults.
Adult females
lay eggs.
Eggs passed in
droppings onto
pasture.
Eggs hatch.
Larvae eaten with grass.
A.C.L.H.

The life cycle of the red worm

organisation has taken place, kicking and biting may occur. Once the horses have settled down they will tend to thrive rather better than they did before.

Size has nothing to do with dominance, the smallest horse quite often becoming the boss.

Where horses are concerned it is imperative that they should be treated as a collection of individuals and never thought of as a group.

For this reason the ideal is to feed every horse separately in a separate loose-box. This enables the individual to eat without free competition from its fellows, but where animals must be fed in a large group, as when they are turned out together in a field, the food should be placed in troughs far enough away, one from the other, to prevent domination of the weaker members in the group. If hay is to be fed, this too should be placed in piles far enough apart to prevent fighting and preferably one more pile than the number of horses present.

Horses should not be fed off the ground, as half the food is often scattered, pawed into the ground and lost by most horses while eating. Also, possibly of more importance is the fact that, unless the pastures are worm free, large numbers of red-worm larvae can be eaten by horses when they crop the grass very short in search of the food.

A suitable feed for the Autumn would be a mixture of oats, bran and stud nuts, fed slightly damp once a day. This can be continued until the end of September or October according to the year in question, when, as the grass begins to die down altogether, two feeds per day and some hay, should be given.

This graph explains the average growth of grass in the British Isles throughout the year. The trough which occurs in mid-Summer is due to drought, when in some areas the grass is completely burnt up and supplementary feeding ,with hay must be carried out, corn should also be supplied at this time to heavily in-foal mares, that is, all except for small ponies which can do well on hay alone at this time of the year. The flush of grass which occurs in September is of lower feeding value, i.e.

higher fibre content and lower in protein, than Spring grass, therefore the supplementary ration should be continued during this period.

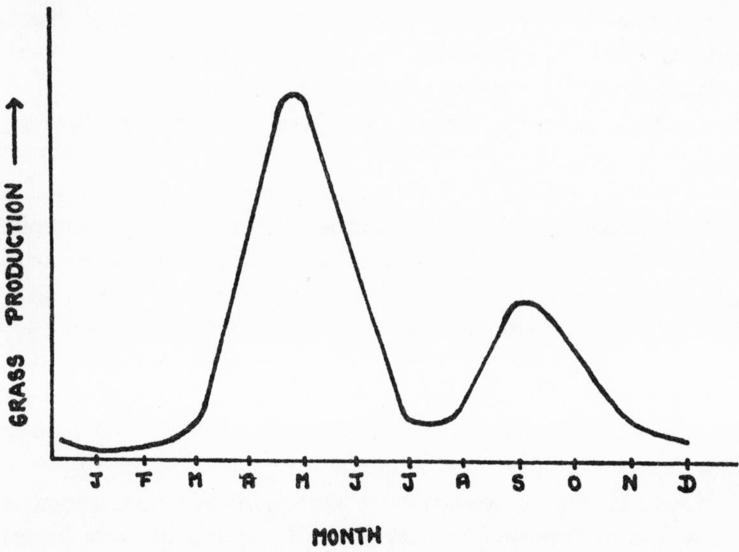

In-foal mares, especially those due to foal early should not be left out at night after the end of October or beginning of November when the weather begins to get cold. They should, however, be turned out or exercised every day up to foaling.

Once the mares are brought in they should have a medium feed of oats, bran and stud nuts in the morning before they are turned out and a similar very small feed on coming in, in the afternoon, and finally a large warm boiled barley and linseed mash at night; this mash can contain any vitamin-mineral supplements or worming powders. Ad lib hay and water should be provided to mares and youngstock and there should be a mineral lick in each box.

When feeding horses it must be borne in mind that no two horses can be treated exactly alike: the good feeder studies the requirements of each horse as an individual and acts accordingly; although a basic ration can be devised as a foundation and guide to the daily ration of horses, this must then be modified to suit the individual, at all stages. Accordingly, horses which tend to run to fat or get laminitis should have less fattening and heating foods than those which tend to remain thin.

The most common fattening foods for horses are: flaked maize and boiled barley, the former can be fed as a replacement for part of the oat ration at the rate of 2/3 flaked maize to 1 part of oats; barley, however, should always be fed whole, boiled, as the grain is too hard in its natural state and would cause colic if fed raw, unrolled. When boiled with linseed (high protein and oil) and mixed with bran and oats it forms a highly nutritive and palatable feed. The barley and linseed should be soaked in cold water over-night and brought to the boil in the morning (this is essential in order to destroy the poisonous cyanogenitic glucoside in the linseed) and placed in a slow oven in a covered pan for several hours, until the grain opens. Allow one good handful of linseed and about six handfuls of barley per horse.

To check if you are cooking your barley enough examine your horse's droppings and if these contain many whole grains the mash should be boiled for a little longer.

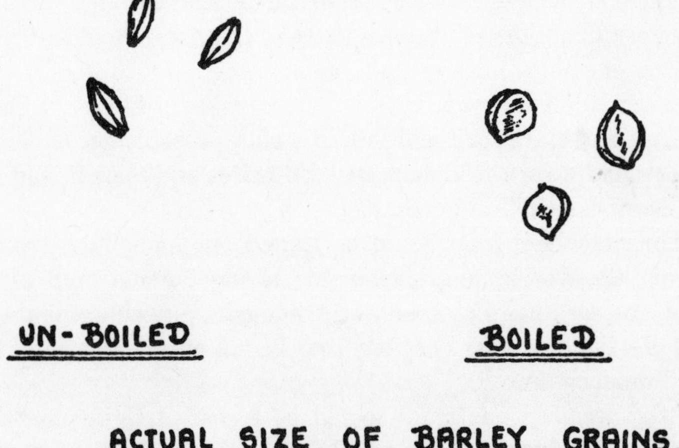

ACTUAL SIZE OF BARLEY GRAINS.

When the mare gets very close to foaling she will probably eat less food due to increasing pressure from the foal, which grows very fast during the last month or so and thus requires more food than ever to keep it going. Consequently, the quality of the ration should be stepped up: less bran should be fed and more stud nuts or oats, only the very best hay should be fed. In order to get the maximum quantity of milk at foaling time, some flaked maize and a boiled barley/linseed mash should be fed every day for about four weeks before the mare is due to foal – but a careful watch should be kept on the mare's udder: if she appears to be coming into milk early the flaked maize should be cut out of the ration.

Immediately after foaling the mare should be offered a warm bran mash made in the usual way: a bucket of bran, some salt and enough boiling water to damp the bran but not make it sloppy, the bucket should be covered with a sack and allowed to cool before feeding.

Young foals should be encouraged to eat, from two days onwards, some warm damp mash will encourage them more than a dry feed. A separate plastic bowl and later a separate manger should be provided for the foal as most mares, however possessive in other ways, will often chase their foals away from the manger, therefore as the foal gets older and becomes capable of eating quantities of feed on its own, the mare should be tied up to a ring at the manger while she is eating.

Except for small ponies, which are capable of living off grass without any extra food and indeed would probably get laminitis if they had a corn feed each day, all mares and foals should be fed every day while at grass.

The mare and foal should be turned out for longer periods during the day in preparation for staying out day and night once the weather becomes warm enough, probably from the end of May onwards. They can then be left out in their paddock until weaning time.

After weaning the mare should be turned straight out onto the barest paddock available and should not receive any supplementary ration until her milk supply has dried up completely.

The foal must be kept in a loose-box for about 3–4 days before he can be let out again and then he should come in every night for safety. This can be continued right throughout the winter providing the weather is suitable.

At weaning time it is of paramount importance that the foal should be kept going and receive as little set back as possible. To this end he should be fed three times a day, possibly increasing to four times a day, i.e. a medium feed in the morning, either one small feed on coming in from the paddock or a little less in the morning and an extra feed at about six o'clock followed by the evening feed at ten o'clock or later, making four feeds in all. If on the other hand only three feeds are given, the last and largest feed can be given at about six o'clock. Horses always do best if fed little and often due to the relatively small size of their stomachs.

At weaning time the accent should be on a high protein ration, i.e. about 18% protein. This can be achieved by adding a high quality calf rearing milk powder to the feed (or the specially prepared milk powder for foals which is now on the market), this can be mixed with raw eggs if these are available, with some glucose added and the whole made up to about a pint with warm water and then poured over the feed and mixed in well. This diet should be continued right through the winter period until turning out time in the spring. If your mare happens to produce twins and a foster mother cannot be found the above ration should be made available to the foals in small quantities from one day old onwards.

When the yearlings first go out day and night onto spring grass, if the quality of grass is available they will not require any additional feed but as soon as the quantity or quality of the grass deteriorates a supplementary ration should be provided, especially if you are preparing them for the show ring or the sales.

During their second winter all youngstock except for ponies should come in at the usual time in the autumn but native pony youngstock can lie out if they are provided with some form of shelter and adequate food. It should, however, be realised that any animal living out in all weather conditions requires extra feed to keep himself warm, as well as that necessary to maintain his condition and for growth.

During the second and subsequent winters a lower protein diet can be fed, i.e. 14–15% protein, until the horse has stopped growing; when he should be fed in the usual way according to the amount of work he is doing.

7

WEANING

Weaning is the process of separating the mare and foal when the foal is old enough to thrive on a solid diet.

The very earliest date at which this can take place is when the foal is four months old. However, the optimum time for the average mare and foal to be weaned is at six months. There is no definite time that weaning should occur and it will depend largely on the general condition and health of the mare (i.e. pregnant or barren) and the size and maturity of the foal.

If on the other hand the mare is pregnant again and even though she may be fat it would be far better, in the interest of the foetus which she is carrying, to wean her present foal by six months. In the case of an in-foal mare in poor condition, it would be definitely advisable to wean the foal as early as possible, i.e. at four months, as this type of mare would find it very difficult to maintain her unborn foal as well as support her suckling foal. She might then possibly go down further in condition which would predispose to an early abortion. Fat barren mares can be left with their foals until the following Spring should the owner so desire.

If the foal at foot is in any way weak or ill, it must on no account be weaned until it has recovered, as the act of weaning is always a shock to the foal's system, and this would undoubtedly set the foal back further. If a sick foal goes off suck the mare must be milked right out at least six times a day, to keep her in milk until the foal has recovered.

The actual process of weaning is not difficult and only requires a little planning. The weaned foal should be put in a roomy loose-box and the mare turned straight out into a bare but well-fenced paddock. The paddock chosen must be out of earshot of the weaned foal and should be reasonably bare to encourage the mare to dry off. All fences should of course be double checked for gaps and weak places before the mare is turned loose.

The loose-box to be used for the foal must if possible be large enough to allow him to have some exercise. The water bucket should be placed off the ground as the foal will probably rush round the box at first and knock any buckets over which happen to be standing on the floor. For the same reason the hay should preferably be placed off the ground, in a manger, hay rack or in a hay net.

Plenty of bedding should be put in the box to prevent injury. As weaning usually takes place in the late summer or early autumn when the weather may be hot, a wire cage should be put over the top door so that this can be left open all the time with no danger of the occupant jumping out. If you have not got a wire cage for your loose-boxes, one can be made quite simply, by making a rough wooden frame, the same size as the top door of the loose-box and nailing some half inch wire netting over the frame, this can then be tied in place with string to the hinges and fittings of the top door.

If the foal is not already receiving a daily corn ration, one must be given for about two weeks before weaning, to accustom the foal to eating corn, before the mare is taken away, otherwise he might receive a very severe setback. On no account should weaning take place until the foal is eating corn readily.

On the selected day for weaning, bed the loose-box down well, put in a little hay, a bucket of water and a mash or dry feed according to what the foal has been used to eating. Bring the mare and foal in from the paddock and take them into the prepared loose-box, put a bridle on the mare with a long leading rein or lunging rein attached. With the help of a friend walk the mare out of the box, leaving the foal inside and shut *both* the top and bottom doors. Ask your helper to walk behind the mare with a stick to prevent her from stopping or trying to run back to her foal as you take her to the paddock.

If at all possible have a quiet horse ready to turn out with the mare for company, let the mare and her companion go and stand by the paddock gate until the mare has settled down completely, which may take several hours. Once the mare has settled and is grazing peacefully it is quite all right to return home and see to the weaned foal.

Some people advocate weaning two foals at the same time and putting them together in the same box. This method, as I see it, has one advantage and two disadvantages:— The two foals are companions one for the other and consequently they possibly tend to settle down quicker and miss their mothers less. It does entail a further period of weaning later on, when the two foals have to be separated. The other disadvantage is that one foal is bound to become the "boss" and unless all the food is provided ad lib, or a constant guard is mounted at feeding time each day, one foal is going to get less than the other to eat. The author has tried ad lib feeding of stud cubes to foals and yearlings from weaning time onwards, with remarkably good results – these youngsters did not ever over eat, but helped themselves whenever they felt like some food (i.e. little and often). The manger was completely emptied at least twice a week to prevent any stale food accumulating in the bottom. This is probably a very good method to adopt if you have to go out to work every day during the week and cannot return home at lunch-time to give a mid-day feed.

When you get home after the mares have settled down and are grazing quietly, if it is a hot day the top door of the loose-box should be opened and the wire cage placed over the top door.

When the foal is mucked out the door *must* be kept closed

and the dirty straw piled against it, then it can be opened just wide enough to allow for the removal of the manure. The top door or wire cage must be kept closed all the time otherwise the foal might quite probably jump out, while you are at the back of the box. Foals will usually remain quietly in their boxes after weaning until they are turned out again each morning, when they may become excited but this is only high spirits and something to be expected in young horses.

After about two to three days after weaning, foals may be turned out for exercise into a well fenced paddock. They must be brought in at night for at least two to three weeks. If you decide to turn them out again all the time a dry warm night must be chosen for the first night out, otherwise there would be a possible chance of the young horses getting a chill. Personally, I never leave my weaned foals out at night until they are turned out in the spring as yearlings. If there is only one foal a suitably quiet companion should be found or the foal should only be turned out for short periods for exercise. A watch being kept on it the whole time it is out, as foals are completely without any fear or knowledge of their own capabilities. This can prove pretty lethal, as they will take on fences, hedges and large gates which would stop any self-respecting adult horse!

When choosing companions for brood-mares and in particular for youngstock, apart from being quiet the companion must be in good health and free from worms. This latter is particularly important with regard to donkeys. Nearly all donkeys are heavily infected with lungworm – *Dictyocaulus Arnfieldi.* This is peculiar in as much as the normal symptoms of lungworm infection are usually absent in the infected donkey but the infected donkey is capable of transmitting its worms to any horse with which it is turned out. The symptoms of lungworm infection are usually a chronic cough with no other evidence of illness or disease; as the condition progresses this cough is usually accompanied by symptoms of 'broken-wind', i.e. the double expiratory lift.

The life cycle of the lungworm is described in the diagram

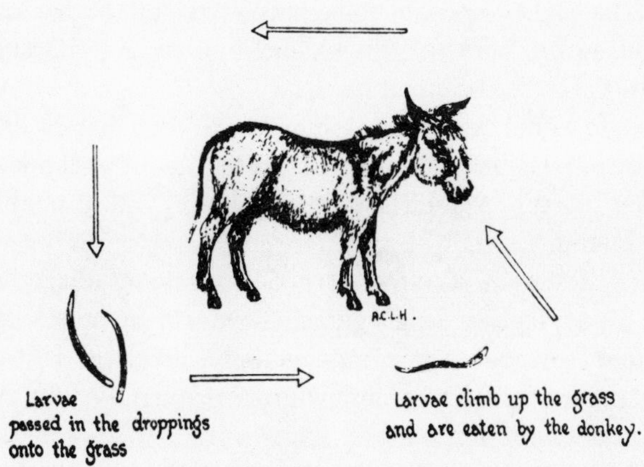
A.C.L.H.
Larvae
passed in the droppings
onto the grass
Larvae climb up the grass
and are eaten by the donkey.

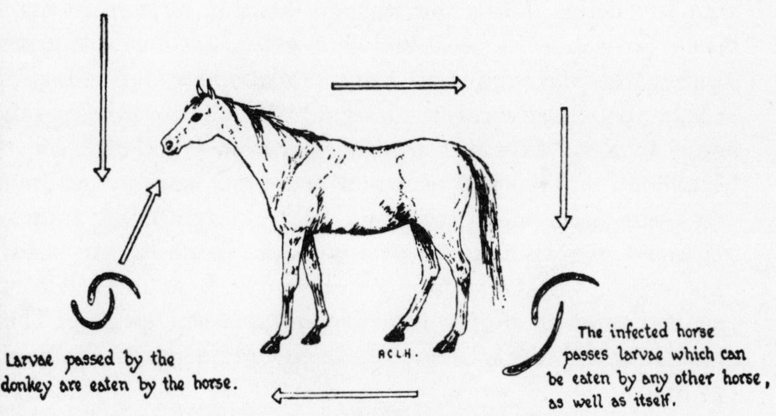
A.C.L.H.
Larvae passed by the
donkey are eaten by the horse.
The infected horse
passes larvae which can
be eaten by any other horse,
as well as itself.

The life cycle of the lungworm

on page 69. On being swallowed the larvae pass down into the intestines, where they penetrate through the intestinal wall and enter the lymphatic system, via which they reach the right heart; from the right ventricle they become blood borne and migrate to the lungs, here they break out into the air passages and eventually reach the bronchi.

There are no other symptoms than a chronic cough and possibly broken wind in the infected horse, unless secondary infection takes place, in which case the horse will probably develop pneumonia.

Lungworm infestation is often very difficult to eradicate as no very safe and effective drug against the worm in horses has yet been found although some are partially effective. If you suspect that your horse has a lungworm infection, veterinary advice should be sought immediately.

It is essential, therefore, to have a donkey's droppings checked for the presence of lungworm larvae before it is turned out with any horses. To do this, enough faeces to fill a half-pound honey jar should be collected and given to your veterinary surgeon who may send them away to a laboratory for testing.

Many people advocate drawing a little milk from the mare's udder for a few days after weaning but from experience, it has been found that this practice tends to cause mastitis rather than prevent it. Very much better results are obtained by turning mares out to grass and just leaving them to walk off the excess milk. Nature works on a supply and demand basis and will very rapidly cut off the supply if the demand is no longer there. The great thing is to allow plenty of exercise and a reasonably bare pasture.

The mare's udder should be checked each day after weaning for any indication of mastitis. If this should happen the udder will become hot and painful to the touch and appear swollen and hard. If a little milk is drawn off it will be seen to be thin and contain clots, the mare herself will in most cases have a temperature and will probably appear dull and listless. The condition must be treated immediately because if neglected permanent damage to the mammary tissue may occur with subsequent loss of udder function on the affected side.

Where several mares are kept, the continental method of weaning may be used. This is now common practice on many large studs in this country and works well.

To operate this method, the mares and foals must be stabled at night before weaning. In order to wean a foal, the mares and foals are led out to the paddocks in the usual way in the morning. When it is the turn of the selected foal, he is led into the paddock with his companions and held while his mother is taken away to a distant paddock where she is turned out with other horses. As soon as she is out of sight, the foal is loosed and soon joins his friends. Foals weaned in this way seldom, if ever, miss their mothers and settle down to graze and play almost immediately.

Loose boxes fitted with metal cages in the walls, so that stabled horses can see their neighbours, are ideal for weaned foals when they are brought in at night.

8

CASTRATION

Castration is the removal of the male sex organs, i.e. the testicles, in order to render the animal more docile and easy to handle. It also removes both the primary and secondary male characteristics, rendering the animal safe to ride and turn out, with mares; in some cases it also makes the animal a better jumper.

The operation must be performed by your veterinary surgeon and normally only takes about half-an-hour with the minimum inconvenience to the animal. It cannot be performed until both testicles have descended into the scrotum.

The horse should be starved the night before the operation.

There are two methods of castration :—

General anaesthetic → Animal on the ground.
Local anaesthetic → Animal standing.

After the operation the horse should be offered a warm bran mash and must be kept under very clean conditions and on no account allowed to lie on dirty bedding, otherwise general septicaemia might result. The horse can be turned out day and night if the weather is warm enough and the animal has been running out immediately before the operation. Exercise is of paramount importance at this time, to prevent swelling and permit drainage of the wound.

Occasionally animals with only one visible testicle are met and more rarely still, those whose testicles never descend into the scrotum. These animals are known as 'rigs' or more correctly as cryptorchids, those with one testicle down being referred to as monorchids.

The castration of cryptorchids may involve a major abdominal operation, as the undescended testicle often lies inside the body cavity.

Uncastrated rigs are usually most undesirable animals; some are capable of getting mares in-foal and exhibit many male characteristics, including screaming and striking with their front legs.

Where valuable pedigree stock are concerned, it is often better to leave them uncastrated until they have proved their worth (or otherwise) on the racecourse or in the show-ring. Nothing is more infuriating than to own a champion gelding, which would have been worth many hundreds of pounds more as a stallion. Therefore, as a rule, all pedigree stock should be left uncastrated until the end of their yearling year or in the case of racehorses until at least the end of their two-year-old season. All other stock should be castrated as foals or yearlings.

Castration is better carried out in the spring once the frosty nights have gone and before the flies get bad or in the autumn when the flies have gone and the frost has not started. In this way there is less chance of infection and the horse may be turned out day and night, so that he can have plenty of exercise. Exercise is of great importance where newly castrated animals are concerned as it reduces the chance of swelling in the hind legs and scrotum to a minimum. Should these areas start to swell unduly, your veterinary surgeon should be called in as some infection may be present.

9

THE HANDLING OF FOALS AND YOUNGSTOCK

For the first year of its life the foal grows very fast and soon becomes enormously strong and playful, it is therefore absolutely essential that it should be handled and halter broken from a day old onwards, if a struggle with a big foal is to be avoided later on.

Time must be found each day to talk to and handle the new arrival until it will approach you with confidence and let you run your hands over it without showing any signs of fear. As far as halter breaking is concerned, for the first few days it is quite sufficient simply to push it round the mare. To do this, place one hand round its hindquarters and your other hand round its chest, so that you can steer it anywhere.

As soon as possible, a first size foal headcollar should be put on – these are preferable to foal slips which tend to ride back up a foal's neck as well as twisting round into its eyes, but do make sure that the headcollar fits well as a foal's hooves are so tiny they can easily slip through a loosely-fitting strap. The headcollar is better put on and taken off each day – in this way the foal will soon get used to being handled around its head; if, however, you prefer to leave the headcollar on all the time, it should be removed and cleaned at least once a week, as a foal's skin is very soft and can easily get chaffed by stiff or dirty tack, also if the headcollar is removed for cleaning every week any adjustment for size can be made, as foals soon outgrow their headcollars, and they will then cut into their skin.

Before you take your animals out of the loose-box (or from the paddock if your foal has been born outside), it should be halter broken, as most foals will run round the mare in wide circles while she is being led and in this way they can literally 'run into trouble', especially if there are any sharp objects such as harrows or barbed wire lying around. Most foals will run backwards and fling themselves on to the ground the first time they feel a pull on the headcollar, so a soft landing is essential.

Wherever possible, care should be taken to prevent the foal flinging itself down, as an awkward fall could possibly cause damage to the spinal column, which might result in wobbler symptoms – inco-ordination of the hindquarters, which is incurable.

Attach a leading rein, which is at least eight feet long, to the headcollar, keep one hand firmly round the foal's hindquarters to push it forward when necessary; your other hand, holding the rein, can be placed round the foal's chest to prevent it from rushing forward. Once the foal is moving forward well, your hands can be removed and he will then be led in the usual way.

Should the foal fling itself over backwards, allow the rein to go slack immediately and don't apply any tension to it until the foal is on its feet once more, the moment it jumps up it will undoubtedly run to the furthest end of the rein but with a little patience you can get your hand round its hindquarters and start pushing it forward again.

If you ever wish to lift your foal onto his feet straighten his front legs out and then place one hand round his tail and your other hand round his chest and lift; never try to lift him by clasping both your hands under his tummy, as you could damage his internal organs and ribs.

When you venture outside the loose-box with your mare and foal, ideally the mare should be led in front with the foal close to her near side. Should the foal get left behind, the mare must be stopped immediately and the foal allowed to catch up, otherwise the mare will become very upset. Some mares, however, cannot bear to go anywhere without being able to see their foals all the time, in which case the foal should be led in front of the mare – this is often especially true when the mare is asked to go into or out of her loose-box, trailer or horse-box.

Once the foal is leading well from the nearside, it is a very good plan to practise leading him from the offside too, as this is very useful, especially if you are ever short of help and need to lead the mare and foal one in each hand, in order to bring them in or turn them out.

During the first fortnight of his life it is advisable to start picking up the foal's feet, as the sooner he gets used to being handled the less trouble you will have later on, when he gets bigger and stronger. Whenever your blacksmith comes to trim your mare's feet, ask him to go round your foal's feet too even if there is nothing for him to do; the foal will then get used to the blacksmith from an early age. It should, however, be noted that the young horse is growing at a fantastic rate during the first two years of his life, so any neglect of his feet at this time above any other, can lead to permanent twisting and damage to his legs in a very short space of time. Therefore, a young horse's feet must be seen to at least once a month regularly: always remember the old adage 'no foot, no horse'!

When your foal is still a baby, on no account be tempted to play with him, most especially if he is a colt: if you allow him to chew and suck your fingers at this age, don't be too surprised if he bites you later on! If you teach him to shake hands or rear, don't be surprised if he strikes with his fore-feet and rears when he gets bigger!

A young foal, like a child has not got his powers of intelligence developed to the same extent as the adult and this must be taken into consideration when dealing with a young horse, but at the same time it is absolutely essential to take a firm line with him and not let him get away with any disobedience, as, if you do, like a child, he will be twice as bad next time. Sometimes it is however better to ignore small offences, otherwise you may find yourself constantly scolding your young horse, which is not good for him or for you, e.g. most colts tend to nip in play and this can up to a point be ignored. Horse psychology, like child psychology, is something which comes with practise.

If your foal is a big one, i.e. the type of strong thoroughbred or hunter foal which will make over 16 h.h when fully grown, it is advisable to mouth him about Christmas time. Otherwise as your foal gets bigger and stronger there will come a day when you will find yourself unable to stop him with a headcollar alone. The young horse should never learn the dangerous habit of setting his neck and charging, as once learnt this is a very difficult trick to eradicate.

In order to mouth your horse and get him accustomed to having a bit in his mouth you will need :

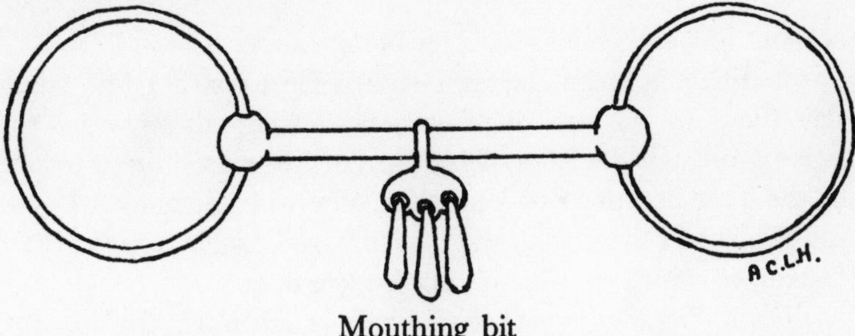

Mouthing bit

(a) A straight bar mouthing bit with keys.
(b) A sound bridle fitted throughout with buckles or studs, of the correct size but with a larger brow-band for easy fitting and no nose-band.

To put the bridle on for the first few times, remove the bit, undo the cheek-piece on the near-side of the bridle and slide the browband off, so that it is only attached to the off-side of the headpiece. Then pass the headpiece over the poll and hold it on the nearside, pick up the brow-band and pass it across the front of the forehead and slip it on to the headpiece in the usual way, the throat lash can then be buckled allowing at least the width of three fingers between it and the cheek. It is advisable not to attach the bit to the bridle before it is put on, as the young horse will often throw his head round while you are putting on the bridle for the first few times and in so doing could give himself a bruise on the side of the face with the loose bit! Therefore put the bit on separately after the bridle is in place: first attach one ring to the far side of the bridle in the usual way – then pass the bit across below the horse's teeth and insert your thumb into the corner of his mouth on the nearside, open his mouth gently and slip the bit in, securing it to the bridle on the nearside; never try to force the bit between his teeth.

For young horses it is preferable to adjust the bit rather too high in their mouths, i.e. so that it wrinkles the corners of the mouth and in this way prevents them from learning the habit of getting their tongues over the bit – especially during the stage of mouthing and playing with the bit and keys.

Mouthing must be carried out in a loose-box but first check that there are no nails or projections sticking out on which the horse can catch the rings of his bit. For this reason the top door of the loose-box must be kept shut, otherwise the horse will put his head over the door and possibly get caught on the door fastenings, which could result in a broken neck.

For the first day the bit should only be left in for ten minutes, so that the horse's mouth does not become sore, for the second day the bit can be left in for half-an-hour and thereafter for half-an-hour to an hour twice a day for two weeks, at the end of which time the young horse should be ready to lead out in hand.

When I come to lead a young horse from a bit I always use a straight bar flexible rubber one:— the type which is constructed on a chain and can be bent up double, not a vulcanite or nylon bit, as being soft I feel the flexible rubber bit is far less likely to do any damage to the horse's mouth, should he play up and you have to take a pull at him. Never attempt to lead a horse off the bit until it is completely obedient to your voice, otherwise you will find yourself having to pull on its mouth all the time.

The leading rein which should be at least 12 feet long, can be attached to the bit in one of two ways:

(a) By using a bit coupling.

(b) By buckling the rein on to the offside ring of the bit and simply passing it across the back of the jaw and through the nearside ring to the leader's hand. In this way the rein rests lightly behind the jaw until the horse requires restraining, when it is pulled tight.

At this stage the first essential for any well schooled riding horse can be taught, namely: Controlled Forward Impulsion; the other two major principles: Correct Bend in All Movements and Even Rhythm at All Paces, will of necessity come later. However, to deal with the first principle – the natural instinct of all horses is to move forward and the young horse must be encouraged to move forward freely in front of you, so that you can take up a position by his shoulder with your right hand about a foot or more from the bit and your left hand holding the loop and slack of the rein – never wrap the rein round either hand, especially when leading from a head-collar, as in the event of your horse taking fright and bolting, the rein would tighten on your hand so you would be unable to let go and would be dragged.

In order to create controlled forward impulsion, you have two aids at your disposal:

(a) Your voice.

(b) Your whip.

The latter must be thought of and used solely as an extension to your arm, every horse will do more for the man he loves than for the man he fears.

The whip should be carried in your left hand and used on the flank or quarters at the same time as your voice, to encourage the horse to walk on. It can also form a useful aid when turning the horse away from you, in which case the handle end should be held up to the horse's face to get him to bend his head and neck to the right – never ever hit him on the face.

Some time before you start to lead your horse out in hand and school him, decide exactly which words you intend to use and always stick to them: Make your words of command short and simple so that your horse can learn to understand them quickly, be ready to make much of him when he does well and always finish a lesson on a good note: i.e. the moment he has understood and obeyed you. In this way he will return to his paddock or loose-box in a happy frame of mind and will tend to do better the next day. Never ever repeat an exercise the horse has just done well otherwise he will probably do it badly again and you will be 'back to square one'; horses always remember a lesson better if they are allowed to sleep on it! With young horses in particular it is essential to keep your lessons short.

As well as being taught to walk and halt in hand, the young horse must also be taught to trot and rein back – the latter should be done by pressure on his chest and not by hauling on his mouth, only one or possibly two steps backwards must ever be attempted, accompanied by the word: 'back', the horse should then be asked to walk on again immediately, to prevent him from learning the habit of running backwards.

Once the horse is one year old, he can be introduced to the lunge: this is the next step in his education after being led and follows on in a perfectly natural order from it.

The degree of maturity of the young horse as well as its conformation, should be considered before lunging is started: an immature horse, one in poor condition or with weak hocks, must be left until it is much older and better able to withstand the work, otherwise irrevocable damage may be done to it's limbs.

The necessary equipment for lunging is:

(a) Lunging cavesson;
(b) Lunging rein;
(c) Lunging whip;
(d) Four Brushing or Polo boots.

The cavesson should be as light and strong as possible with a felt or well-padded noseband – unlike an ordinary bridle, the nose band and throat lash straps are done up as tightly as possible; this is because a loose noseband will chaff the nose and unless the throat lash is tight the cavesson will work round into the horse's eyes.

The lunging rein is attached to the front ring on the cavesson and should be approximately 25 feet long. When buying a lunging whip the very lightest one possible should be chosen, as these are less tiring to use.

It is essential that the horse and a young horse in particular, should have some form of protection on his legs when he is lunged, to prevent damage to them should he knock himself. For this reason brushing or polo boots should be worn, or alternatively gamgee tissue or sponge rubber sheets can be applied to the legs under an ordinary exercise bandage.

From my own experience I have found it advisable not to use side-reins when lunging a young horse as they tend to stiffen him and apart from learning obedience the object of lunging is to obtain a correct bend and to cultivate free forward movement. Should the horse try to stiffen and turn out of the circle, perhaps towards his loose-box or the paddock gate it is far easier to pull his head round quickly without side-reins than it is with them.

To teach your horse to go on the lunge, put the lunging tack on as described above and lead him out to the paddock. Using the words of command you have already taught him, lead him round in a circle to the left but gradually let him have a longer rein and if your original lessons were a success you should find that your young horse will walk round you in a circle on the lunge without any trouble. If you get into difficulties, ask a knowledgeable friend to help you, by leading the horse round.

Continue to walk in a small circle yourself until your horse is really balanced – this can take several months, after which time you can stand still in the middle of the circle and let the horse go round you; grind one heel into the ground and use this as a pivot, change your pivot foot when you change the horse's direction. When lunging keep level with a point just behind the horse's shoulder and keep your whip level with his hindquarters; this is still used as an extension of your arm to keep the horse moving forward.

It is essential that you should change rein every few minutes when lunging otherwise the horse will tend to become one-sided if lunged continuously in only one direction.

To change rein on the lunge, first halt your horse and then walk up to him slowly, taking up the slack of the rein as you go, talk to the horse as you approach him and then make much of him, walk round to his other side – slowly transfer the rein into your other hand as you change side and at the same time change your whip into your other hand by passing it across behind you and away from the horse. Tell your horse to walk on, and start by lunging on a small circle and gradually increase the size as the horse settles down again.

When the horse first goes out on the lunge in a morning he will probably shoot off, bucking like mad. In this case you must just stand in the middle of the circle and wait until he has settled down before you ask for any obedience, as in this frame of mind the horse would be bound to ignore you.

Any words of command can be used as long as they are clear and intelligible to the horse and always with the same intonation of voice. The words of command I use for my horses are:

For increases of pace –

Walk on; T – rot (with a long T); Can – ter (in two syllables).

For decreases of pace –

Walk (long drawn out); and a long drawn out Whoa for the halt.

Should a slowing down word of command ever be ignored, repeat the word and give a sharp jerk on the lunging rein: the action is transferred through the tightly fitting noseband of the cavesson to the horse's nose and usually slows him down immediately. Therefore a horse must never be lunged directly off the bit as a similar action would damage the bars of the horse's mouth. If an order to increase the pace goes ignored, crack the whip behind the horse but never actually hit him. To prevent a horse from cutting in on the circle, move your lunging whip across in front of you until it is level with the horse's shoulder.

It is very necessary to teach your young horses to go correctly on the lunge, as this is a good means of exercising horses when the fields are too wet in the winter or early spring to turn them out.

10

PART ONE
SHOWING IN HAND

If you are going to show your mare and foal or youngster in-hand, it is essential that you should decide exactly how many shows you intend to enter right at the beginning of the season and stick to this number, otherwise showing can become a very expensive game, even if you happen to win prizes most of the time. It is also essential for the same reason, that you should have your own transport.

It is better to make up your mind early in the year which shows you intend to enter, especially if you are going to go to some of the bigger ones, as their closing date for entries is often very early in the year and many months before the actual show date.

If you are not sure which shows to patronise or indeed where to apply for the schedules, the best thing to do is to buy a copy of the special show number of *Horse and Hound* magazine, which comes out in the first week of March; this will give you a complete list of almost every show to be held during the year, together with the secretaries' names and addresses. It also contains many helpful articles connected with shows and showing as well as a very comprehensive list of stallions at stud in the British Isles.

When you have decided which shows you might like to enter, write to the secretaries concerned and ask them to send you their schedules and the required number of entry forms, i.e.

one for each horse. The schedule will give you a list of classes, judges, prize money and special prizes as well as the shows' rules and regulations.

Once you have completed one season's showing there will be no need to write for any further schedules as these will be sent to you automatically, from the previous year's shows and usually many more schedules besides from other shows which you did not enter in the previous year.

Small shows will very often accept entries on the actual day of the show for an additional entry fee. This must, however, be checked in the schedule as not all shows offer this concession and nothing would be more disappointing than to arrive at the show only to find that you cannot enter.

When filling in your entry form make certain you fill in all the relevant details and sign the form where required. It is also important that the correct entry fees are included, as without these the entry will not be accepted.

At most of the larger shows you will find that you can hire a sleeping cubicle for yourself or your staff. These are sometimes provided with a bed and matress but at other shows they contain nothing. If these are not situated in the horse lines, some people prefer to book an extra loose-box so that they can be near their horses all the time; or the extra loose-box can be used for storing your saddlery and feed in; it is usually safer to have a large box made, which will take all your show equipment and which is fitted with a key, than to leave any equipment in an unlocked loose-box on the show ground.

It is a very good idea to keep a notebook of every show attended, the judge, the horse shown and the prize won – if any; in this way you will be able to see at a glance, in future years, exactly which judge likes which horse or type of horse. This sort of information can be most useful when you make your show entries each year.

If you have followed the schooling programme as suggested in the chapter on handling foals and youngstock there will be no need to carry out any extra schooling sessions especially for showing, as the only requirement for showing in hand is that the horse should be able to lead well at both the walk and trot. A certain degree of high spirits is usually overlooked as far as foals and yearlings are concerned; in point of fact it is far better if they are on their toes, as they then tend to look better in the show ring than something which is very well behaved and at the same time half asleep.

It is, however, an absolute must that any horse which is going to be shown regularly should be good to load into a horse-box or trailer. Otherwise with a reluctant horse at a show you will find yourself surrounded by dozens of people all offering advice and a horse, as the crowd gets larger, becoming more and more determined not to be loaded. Therefore, if you have your own means of transport, it is a very good plan to practise loading and unloading the horse long before the actual show date. To be on the safe side the earlier you get a foal accustomed to being loaded and travelling the better. While the foal is still with its mother and is quite small, loading will present few problems unless the mare is bad to load. Once the foal is weaned it should be taught to load, as soon as possible. To do this I load my foals into a trailer every day and take them up the road to turn them out and bring them back to their loose-boxes in the trailer at night; in this way they get very used to being loaded from an early age, and indeed they will usually walk into the trailer on their own after only a week or two. I do however appreciate that this method would not be very practicable from the point of view of anyone who has a number of weaned foals, as it would obviously take far too long each day to turn everything out, but for the small breeder with only one or two foals each year, it is an admirable method of getting them used to travelling.

For older horses which prove difficult to load, a lunging rein attached to the side of the trailer or horse box and passed round the back of the horse just above the hocks and then slowly pulled up tightly on the far side of the trailer or horse box, will often cause the horse to walk in easily. The horse's head should not, however, be pulled hard, nor should the person leading the horse turn round and look at him. Should the horse rear-up and be in danger of coming over backwards the lunging rein should be dropped immediately. A few oats in a bowl will often entice a horse into a horse-box, or failing this, if the horse is being genuinely stubborn and is definitely not frightened, one or two hard bangs on the rump will often do the trick, but make sure the horse is not frightened first, otherwise you will do more harm than good.

If the horse can be trusted not to kick, an alternative method to the lunging rein is for two people to link hands behind the horse and lift it up the ramp. With a young horse that is not used to picking up its feet to put them on the ramp, it is often a very good idea to do this for him, one at a time; it is also often a good idea to scatter straw down the ramp, to make it look more inviting.

The best time to trim your horse in preparation for the show season is as the spring approaches and your horse starts to lose his winter coat. At this time of the year you will find that the mane and tail hair comes out much easier, as does any surplus hair around the fetlocks and head.

In all breeds where a clean heel denotes a sign of breeding as in Thoroughbreds, Arabs, Hacks, Hunters and Riding Ponies, etc., this is best done a little at a time, otherwise, if there is a lot of hair, the area will get very sore. On no account should the hair be cut with scissors, as this will show as ridges down the fetlock and look terrible. The fetlock should be plucked out completely.

Any surplus hair on the face should either be plucked out if it is loose or it can be singed off with a lighted taper. The tough whiskers around the nostrils and mouth can be trimmed off carefully with a sharp pair of blunt ended scissors.

If you do not intend to show your mare or youngster with his tail plaited, his tail should be pulled. This must also be done over a long period of time, like the fetlocks, only a few hairs taken at once. A tail bandage should be put on the tail all the time that the horse is in his loose-box to encourage the hairs to lie down.

The mane should also be pulled, preferably up to about three or four inches long, so that it will only need pulling once more before the show season is over. Unlike the tail this can be done all at once. The mane should be well brushed out and a flat-topped metal comb should be used. Taking only a few hairs at a time, back comb them to the required length and wrap the longer hairs you are holding in your left hand, round the comb and pull them out, continue down the mane keeping a perfectly straight line along the neck. If the mane is too thick, it can be thinned by pulling hairs out from the underside but never on top. A thin mane can be shortened by using a sharp knife instead of pulling the hairs out by the roots after you have back-combed the hair, the hairs should be broken across the length of the blade as this gives a more natural appearance. On no account should a mane ever be cut with a pair of scissors.

All horses' feet should be trimmed regularly – at least once every six weeks, but where you intend to show a horse its feet should not be touched preferably for about two weeks before the show otherwise if the ground happens to become hard, your horse may go foot-sore if he has been newly trimmed. Some people always put light shoes on the front feet of all the young horses and brood mares that they show. Minor faults in the front feet can sometimes be rectified by the use of shoes but from my own experience I have always found that more horses move better without any shoes on, than do with shoes on, although there are of course exceptions to this rule, these are mainly horses with split walls or thin soles.

Horses which turn their toes in or out can have this fault rectified by careful shoeing from an early age. The outside of the shoe is often built up to throw a toe out, and built up on the inside to bring a toe in. An older horse with pin toes – toes which turn in, can be made to look straighter if the clips on the shoes are off-set to the outside slightly, while if the horse has front feet which turn out, clips placed slightly to the inside of front, will also tend to make the horse's feet look straighter.

When preparing horses for the show ring, especially if these horses are youngsters out at grass all the time, their coats must be kept in very good condition so that a surface shine can be obtained at a moment's notice without necessitating the removal of any grease from the coat. This is best done by feeding linseed and seaweed meal in the ration each day – as described in the chapter on feeding. Show horses should be fed every day even if they are out at grass day and night. To get them into the required condition for the show ring, about half a two gallon bucket of mash per horse, per day, is usually sufficient for horses kept at grass all the time, until the quality of the grass decreases to such an extent that additional feeding becomes necessary. Small ponies are probably far better without any extra feed at grass, except for hay, as they are usually prone to attacks of Laminitis especially if they receive any corn.

When horses are shown straight off grass they are inclined to "dry up" on coming inside – even for one night. It is, therefore essential to feed cut grass as well as hay when you bring these horses in for a show, otherwise in only a matter of hours they will appear really hollow flanked and run up.

Very little special equipment is needed for showing in hand, the main requirement is that any bridles, headcollars or leading reins taken to a show should be spotlessly clean and shining. Brood mares can be shown in special in-hand showing bridles or double bridles – the latter is more usual when showing larger ponies, Hunter or Thoroughbred brood mares. All show bridles must be fitted with nose bands, otherwise the horse's face will appear bare and unfurnished.

Double bridles used for showing should preferably be stitched or fitted with studs rather than buckles – as buckles tend to look clumsy. All the metal parts of the bridle must be clean and shining: this includes any buckles as well as the bits. Stainless metal buckles are the easiest to keep clean, but any rust can be removed from other types of buckle by rubbing them with very fine sand paper.

For ease of work and first-class appearance there is nothing to beat stainless steel for all bits. Although nickel can look nice, it does require an enormous amount of work to keep it up to show standard – without any mark or stain.

Foals should be shown in brass-mounted headcollars fitted with narrow brow-bands. These headcollars are usually made from leather but any other serviceable material would be acceptable – some people like to use white webbing headcollars for very young foals; these look nice but have the disadvantage of having to be cleaned very regularly with white canvas shoe cleaner which might have a disastrous result if it rained! Brass mountings do, however, look far better than tin, but all the brass work must be polished and shining. Quiet yearling fillies may also be shown in headcollars only, without bits, but entire colts, geldings and older fillies are all safer with bits in their mouths. For leading young horses in-hand, as already explained, I prefer to use a flexible rubber bit and this can, quite satisfactorily be attached, by means of special leather bit attachments (which can be obtained from any saddler) to the headcollar, or better still they can be put on to a special brass-mounted in-hand bridle. Either method is equally correct but the in-hand bridle possibly looks a little neater (see photograph).

As far as I am concerned, I always clean my show tack on the top side of the leather, with dark tan stain boot polish and apply saddle soap to the reverse side only. After a few applications of polish, the leather begins to adopt a lovely reddish tone, which looks well on most horses.

An extra shine can always be obtained by putting the polish on in thin layers and rubbing small areas at a time up to a shine with a little water. A glass-like surface can be produced in this manner, which is useful for the front of nosebands, brow-bands and also for the toes and heels of jodhpur boots. Great care must however be taken in not producing this extra shine too often, otherwise the leather may eventually crack.

As far as possible, oil should not be put on show tack, unless it has become very hard, otherwise the leather will become dark and it will be impossible to obtain a shine on the top side of the leather.

There are two types of leading rein in common use at shows, these are leather leading reins and white web leading reins.

Leather leading reins should never be cleaned entirely with saddle soap, otherwise if it rains the leather will become slippery and should the horse take a pull, the rein will simply slide straight through your hand. These reins are the best and most reliable type, as they are unlikely to become rotten and break suddenly, as do web leading reins, but they should be cleaned in the same way as the show bridles, with saddle soap applied to the reverse side of the leather only, and boot polish to the top side. As already explained the saddle soap should only be applied very thinly to the reverse side, otherwise the rein may become slippery.

White web leading reins are cheaper to buy than the leather variety but they last for a much shorter time, probably due to the fact that they have to be cleaned with white canvas shoe cleaner every time they are used and this tends to rot them eventually.

The leading rein can be attached to the bit in one of two ways: either it may be buckled to the far ring of the bit and passed through the nearside bit ring to the leader's hand; this method probably gives the greatest control but is consequently harder on the horse. The other method is to use a bit coupling, which is usually made from leather and simply buckles or clips onto the rings of the bit on each side and has a brass ring in the middle to which the leading rein is attached; this method produces a more even pressure on both sides of the horse's mouth. The leading rein can also be made with a bit coupling already attached, but make sure that the horse cannot get hold of the bit coupling with his teeth, otherwise it may get caught on his bottom jaw and cause him to rear up and come over backwards. If your horse is very quiet but you still wish to show him with a bit in his mouth, which gives a more finished appearance, the leading rein may be attached to the nose-band of the bridle and the horse led from this entirely, or it may be attached to both the nose-band and bit coupling, to lessen the effect of any pull on the horse's mouth and transfer some of the pull to the nose-band.

The appearance of a beautifully turned out horse can be absolutely spoilt by a badly turned out leader, which after hours of work on the horse is a great pity to say the least.

For showing in-hand where it is essential that you should be able to run fast when the occasion warrants, it is essential to wear a pair of shoes which are easy and comfortable to run in – long riding boots and high heeled shoes are unsuitable. Jodhpur boots look smart when worn with fawn slacks or flat shoes with a skirt.

Men usually wear a neat country style suit or sports jacket and cavalry twill trousers with collar and tie and usually a bowler hat. In warm weather, a cotton jacket can be worn, but however warm the weather is, it is simply not done to appear in the show ring in your shirt sleeves.

Ladies can either appear in the ring in a skirt or they can wear a sports jacket and fawn or light coloured slacks, with a white shirt and tie. Many people wear hard hats, bowlers or head scarves when showing horses in hand.

A Complete List of Articles You Will Need to Take to a Show if You Intend to Stay Overnight

Headcollar
Rope
Hay Net
Water Bucket
Feed
Hay
Feed Bucket
Scoop
Mucking Out Fork
Roller and Tail Guard
Knee Pads

Grooming Kit:
- Body Brush
- Dandy Brush
- Water Brush
- Curry Comb
- Mane Comb
- Hoof Pick
- Stable Rubbers

Show Bridle
Lead Rein

Leg Bandages and Gamgee Tissue
Tail Bandages

Lunging Cavesson
Lunging Whip
Lunging Rein

Plaiting Equipment:
- Thread
- Blunt Needles
- Scissors
- Box to Stand on

Stick

Dog Chalk Block
Hoof Oil
Rag
Vaseline
Disinfectant Spray
Fly Spray
Boot Polish
Saddle Soap
Cloths and Sponges
Brass Cleaner

Wire Cage or Rails – for the front of the loose-box, if necessary.
Show Tickets.

PART TWO

When you arrive in the vicinity of the show-ground, you will find route signs indicating Livestock, Horses, etc., which will direct you to the correct entry gate. Be prepared to show your tickets to the man standing at the entrance gate and make sure that any livestock passes are signed where necessary. Many of the larger shows have a veterinary surgeon on the gate to check all the horses as they come in to make sure that no horse entering the show-ground has any contagious disease, such as a cough or runny nose.

Once your horse has passed the veterinary surgeon, make your way to the horse lines where you should find the horse foreman. He will tell you exactly where your loose-boxes are, probably he will also have the ring numbers. At most shows you can drive down between the rows of boxes to unload your horses, which saves walking them from the horse box parking ground.

When you have found your loose-boxes, check that the manger is clean – if it still contains stale food from a previous horse this must first be cleaned out and the manger and walls of the box sprayed with a disinfectant spray in case the previous occupant had any disease. Straw is provided by nearly all shows so if the box has not by any chance been bedded down, sufficient straw should be asked for. When the boxes have a canvas sheet across their fronts, this should be tied up with string or baler twine before the horse is unloaded and asked to walk into the box, otherwise it will flap round his head even if held up and may frighten him. If you have hired an extra box to use as a saddle room or if you have hired a cubicle which you intend to use as a saddle room, you should unload all your saddlery and feed before you move your horse box as this will save a lot of carrying of equipment later on. When everything has been unloaded you can then take your horse-box to the parking ground.

At some shows the loose-boxes have rather insecure door fastenings and it is therefore far safer if the bolt is tied up with string or padlocked before the horse is left, as the bolts on these doors can easily slip open.

Preferably before you park your horse-box, leave the horse with a full hay-net, feed and water. Water taps are usually located at the end of each horse line with a galvanised tank under the tap – always draw your water directly from the tap, never dip your bucket into the tank, as this water is often stale and contaminated by other people's dirty bucket bottoms.

As soon as you have found somewhere to park your horse-box return to the horse lines and make sure your horse has settled down all right.

Should the show provide hay and green-stuff for their live-stock exhibitors, it is better to go to the forage yard as soon as possible as they often close fairly early in the evening and don't open again until early the following morning. Forage tickets are either sent by post with the other show tickets or they are obtained from the horse-foreman. At most shows these days the green-stuff is tipped at the end of the horse lines for exhibitors to help themselves, but hay is only provided on the presentation of tickets at the forage yard.

Before it gets too late, it is a good idea to find the Members' or Stockmen's canteen (the latter is always far cheaper than the former) and satisfy the inner man; unless you intend to eat out in the neighbouring town or do your own cooking. If you don't actually sleep on the show-ground you will miss half the fun and probably fail to make many friends. If you intend to sleep in your horse-box this must be mucked out before it gets dark otherwise the operation will be doubly difficult and the floor tends to dry off quicker in daylight. Also before you turn in for the night make quite sure you know where the show ring is, otherwise you may have to waste valuable time the next morning looking for it.

If you are sleeping in a horse-box or trailer, it is definitely advisable to have an interior light fitted, otherwise you will be groping around in the dark when you go to bed. As far as the actual bed is concerned, either a camp bed can be used or if you intend to sleep on the floor in a sleeping bag or on a mattress, sufficient straw should first be placed on the floor to make a thick mattress, as most horse-boxes and trailers have corrugated floors which are incredibly uncomfortable to lie on for any length of time, but with a thick mattress of straw underneath, they can be quite comfortable.

At most shows the youngstock and breeding classes are scheduled to start at 9.0 a.m., so when you turn in for the night set your alarm clock for about 5.0 a.m., as an early start in the morning is essential.

When you get up in the morning and have washed and dressed, go straight to the horse lines and start by mucking out, the dirty straw should be thrown in a pile outside the box door from where at many shows it will be collected by a gang of men with a tractor and trailer at about 6.30 a.m.

The loose-box should be bedded down well and the straw built up round the sides so that it looks tidy when the general public come around later to look at the horses. If your horse is inclined to get upset by having lots of people looking at him and stroking him or if he is a colt and has a mare or filly in the box next to him it is often advisable, in the case of open fronted boxes, to tie fine mesh wire-netting across the front of the loose-box and if he is a tall horse to put a canvas or wooden screen round the top of the walls inside the box, to prevent him touching noses with the horses in the adjoining boxes.

As soon as the horse has been mucked out, any stable stains and top dust should be removed from his coat and his mane and tail brushed out well. Then if he is to be shown in any class other than mountain and moorland ponies, palominos or pure-bred Arabs, he must be plaited up and this includes Arabs, palominos and mountain and moorland ponies shown in classes other than those listed exclusively for their particular breed or type, e.g. Riding Pony Breeding or Hunter Breeding classes.

To plait a horse's mane and tail you will need the following equipment:

(i) A headcollar and rope: the horse can be tied up or if he is a youngster he would be far better held by someone else.

(ii) If you are small or your horse is tall, you will need a box to stand on.

(iii) A water or dandy brush to damp the mane and tail, and of course, a bucket with some water in it.

(iv) Some thick harness thread (obtainable from most saddlers), as near to the colour of the horse's mane and tail as possible.

(v) A mane comb and scissors.

(vi) At least one blunt needle – also obtainable from most saddlers.

Before you start plaiting your horse's mane, examine the conformation of his neck, a horse with a slightly ewe neck should be plaited up loosely especially over the base of the neck, as this helps to fill up the hollow on the top line of the neck; whereas

The author's mare Chunky Clemantine. Photographs demonstrating the difference a nose-band makes to the appearance of a horse's head. The bridle used here is an in-hand show bridle, sometimes known as a stallion bridle (*see* chapter 10, part 1)

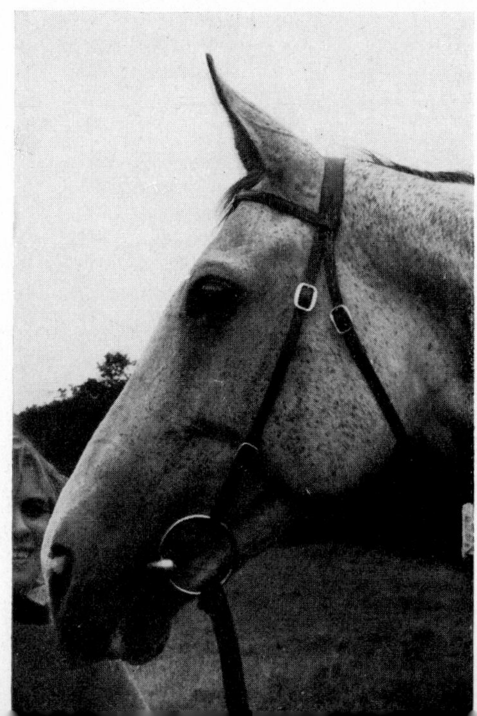

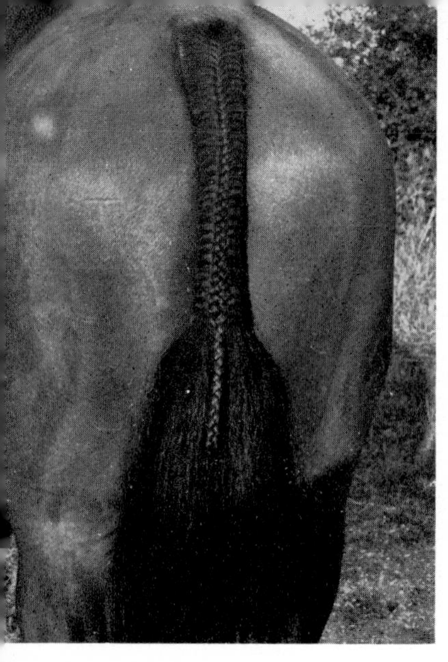

A plaited tail. The secret of good plaiting is only to take a few hairs at a time (*see* chapter 10, part 2)

Stages in plaiting a mane. The first plait is made immediately behind the ears, each plait must be the same width with a straight parting between every one. (1) Plait the hair down to the bottom, secure the end of the plait as described on page 98, any loose hair should be wrapped round the bottom of the plait and secured with a stitch. (2) The long plait is doubled up, and a stitch passed through the double thickness. (3) The thread is passed through the loop of the plait. (4) The plait is doubled up again and secured. (5) The plait is doubled up again, when necessary. (6) The thread is passed round the base of the plait and a stitch passed through the thickness of the plait is secured on the under-side. This is the finished plait. Always use a blunt needle when plaiting-up horses

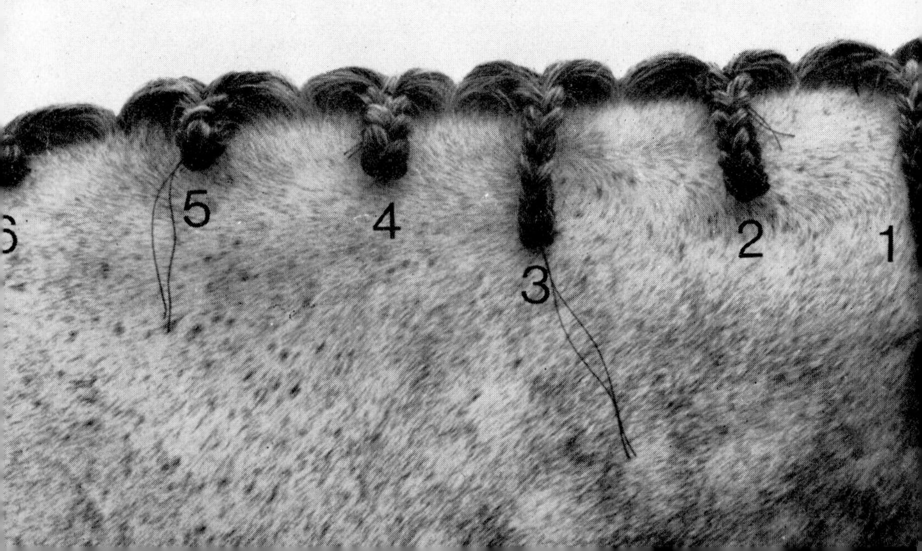

STAGES IN PLAITING A TAIL

1. TAKE A FEW HAIRS FROM THE UNDERSIDE OF THE TOP OF THE TAIL AND PASS THEM ACROSS INTO YOUR LEFT HAND.

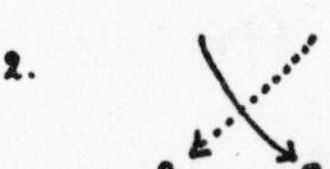

2. TAKE A FEW HAIRS FROM THE SAME PLACE ON THE OPPOSITE SIDE OF THE TAIL AND PASS THESE ACROSS THE FIRST, INTO YOUR RIGHT HAND.

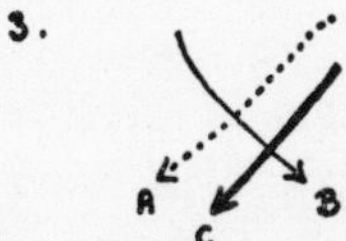

3. INTRODUCE SOME MORE HAIRS FROM THE RIGHT SIDE, IMMEDIATELY BELOW THOSE ALREADY TAKEN FROM THIS SIDE, USING YOUR THUMB AND FIRST FINGER TO HOLD THE CROSSED-OVER HAIRS, PASS THE NEW HAIRS OVER THE TOP OF THE OTHERS.

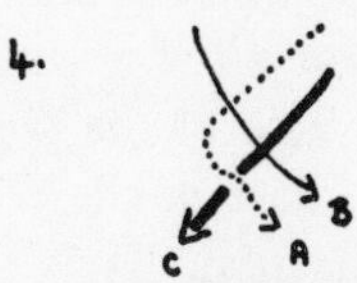

4. BEND HAIRS "A" ROUND OVER THE TOP OF HAIRS "C", PASSING FROM LEFT TO RIGHT.

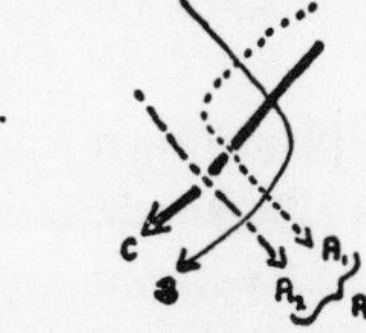

5. INTRODUCE SOME MORE HAIRS FROM THE LEFT HAND SIDE AND ADD THESE TO HAIRS "A", BRING HAIRS "B" ACROSS THE TOP OF HAIRS "A" TO THE LEFT SIDE OF THE TAIL, INTRODUCE SOME MORE HAIRS FROM THE RIGHT TO JOIN "B", AND REPEAT DOWN THE TAIL.

a horse with a short or thick neck will need more plaits put in as tightly as possible to lessen the appearance of the horse's crest – unless he is a stallion, when a crest is usually taken as a sign of fertility in the male horse.

An average of eight plaits are usually put in, more or less as required, the exact number should be decided before you start, then the size of each individual plait can be determined, as the width of hair taken for each plait should be the same to give a neat appearance on the near side of the neck.

Thread the needle and with the thread double tie a knot in the end, stick the needle in your lapel and pass the loose end of the thread over your shoulder so that the needle does not get pulled out, as there is less likelihood of the thread getting caught up in something, if it is hanging down your back.

First damp the mane, divide the forelock from the first plait immediately behind the ear by making a parting with your comb and divide this first plait from the second in the same way, measuring its width with your comb.

Plait the hair right down to the very bottom and secure the end with a knot as shown in the photograph, then bring the needle back into the base of the plait pulling the end up to form a loop, bring the needle back into the end of the loop and double the plait up again, make sure that any stitches on the surface of the plait are kept small, so that they don't show, next wrap the thread two or three times round the base of the plait, pass a stitch right through the thickness at the base of the plait making sure that you catch every layer of the plait, knot the thread off on the underside and cut the thread short enough so that it doesn't show under the plait.

Keep the mane damp and continue to plait down to the withers, the last plait or even two plaits can be made a little slacker to allow the horse to stretch his neck out without it pulling too much. Once the mane has been plaited up completely any loose hairs which are sticking up can be pulled out – on no account should the hairs be cut otherwise in a week or two they will have a hedgehog appearance, and look ghastly.

If your horse's tail is not pulled: when you have finished plaiting the mane turn the horse round with his back to the front of the box, make quite sure that there are no knots in the tail, then damp the top of the tail and start plaiting as high up the tail as possible: take up three strands at the top of the tail, introduce a further strand from the right side underneath the tail and then from the left side underneath the tail; see diagram on page 97. Continue down the tail until you are about an inch above the end of the dock – unless the horse has a very short tail, as in the case of foals and some yearlings, when a considerably shorter length of plait should be made. When you get to the required length, stop introducing hairs from the underside of the tail and simply plait the hairs in your hands down to the end to form a long pigtail, the end should be secured in the same way as for the mane, the pigtail can then be looped up and its end tucked into the plaited part of the tail and secured with small stitches. The secret of good tail plaiting is to take only a few hairs from the undersurface of the tail each time otherwise the plaiting will not be tight.

NEVER EVER PLAIT UP A FOAL'S MANE, as they have not got the neck development to look nice, a foal's tail should, however, be plaited.

As soon as you have finished plaiting the tail, put on a damp bandage – this helps to flatten the hairs and greatly improves the appearance of the plaiting – when you come to take off the bandage, just before you go down to the collecting ring, be very certain that you do not pull it off in the usual way as this would damage the plaiting, instead the bandage should be very carefully unwrapped.

In the case of horses shown unplaited, the mane and tail should be brushed out thoroughly and then combed through, so that it is completely free from any knots. With Mountain and Moorland ponies shown in hand unplaited, if any small section of the mane defies all efforts to get it to lie down flat, a narrow pigtail of the offending section of the mane can be made – this often occurs immediately behind the ears.

When you have finished plaiting up, refill your horse's water bucket, give him a feed and go and have a wash and brush up yourself before breakfast. Immediately after breakfast any vetting, measuring and collecting of ring numbers can be done. At least one county show requires all its entries to be vetted thoroughly before any ring numbers are issued, as well as checking all horses as they come through the gates. Only licensed stallions are not required for the second vetting. All horses entered in height restricted classes which do not possess a current height certificate must also be measured at most shows and at some shows all brood mares are checked for hereditary unsoundness. All premium winning brood-mares and fillies are also checked for hereditary unsoundness after the classes at all shows, if they have not been previously checked before at that particular show.

Once your horse has passed the veterinary examination, and/or been measured and you have collected the ring numbers, the horse can have his final brush over. If he is normally out at grass all the time, a very satisfactory result can be obtained by running a damp body brush over his coat and as soon as it dries, shining the surface with a coarse stable rubber – this method does not get rid of the natural grease in the horse's coat, which he will need when out at grass, it only shines up the surface of the coat – this is quite sufficient when showing brood-mares and youngstock.

Any white socks or stockings should be rubbed over with a block of dog chalk and the horse's hooves oiled, a little vaseline can with advantage be smeared over the almost hairless region around the eyes and muzzle.

When the horse is quite ready, the person who is going to lead him in the show ring should get washed and changed, and the correct ring number must be tied round his waist – the only time these numbers are put round the horse's neck are for Heavy horses and Carriage horses, e.g. Cleveland Bays, Shires, etc.

The show bridle and rein can then be put on the horse and

the tail bandage removed, wait in your loose-box until the first call is made over the loud speakers or by the Horse foreman, for the entries in your class to go down to the collecting ring and then make your way down slowly, there will be plenty of time. The collecting ring stewards will tell you when to go into the ring for judging. If you have never shown a horse in-hand before it would be better if you allowed one or two seasoned exhibitors to go in front of you and give you a lead, otherwise it is quite a good idea to lead the class in yourself, as the judge is bound to look at the first horse into the ring even if he doesn't like what he sees! Otherwise if it is at all possible place your horse between two obviously inferior ones so that the comparison makes your horse look even better.

Walk out smartly on the right rein, if the horse in front of you is walking slowly go on a slightly wider circle but don't let the horse get obscured from the judge's view by another horse.

The judge will now be assessing his class and he will soon start to pull the horses he likes best into the centre of the ring – keep half an eye on the judge and be ready to come into the middle of the ring should the ring steward ask you to. When you are pulled in stand your horse up facing the grandstand and make sure he is standing with his weight evenly distributed on all four legs. The judge may come up to each horse and examine it while it is in line, otherwise he will only examine it when it is pulled out of line in front of him. Wait until the right steward tells you to bring your horse out for the judge to see him, then stand him preferably sideways on to the crowd and in front of the judge. Make sure he is standing with his weight evenly on both front legs and his hind leg furthest away from the judge slightly forward.

When the judge has finished looking at him he will ask you to walk to the far end of the ring, turn your horse and trot back past him to the other end of the ring – as you go up the ring make sure your horse is really walking out to the best of his ability. As you turn walk round your horse, never ask the horse to walk round you otherwise he will tend to become unbalanced and throw his legs out. The horse must also be balanced when he is asked to trot on, therefore when you turn to trot back up the ring walk your horse for at least three strides before giving him the command to trot on. Be ready to give him a tap with your stick if he fails to 'get the message'.

Providing you have a horse which moves really well, in the author's opinion, it is better to trot out fast and risk your horse breaking into a canter than to jog up slowly, in this way the judge can see before your horse breaks that he is an outstanding mover. Once your horse is trotting on well (some horses seem to 'change gear' as they lengthen their stride) drop your hand so that the horse is going on a slack rein, in this way he will have to carry himself and not lean on the bit, with his weight on his fore-hand and his hind legs trailing behind him. Some Arab and pony judges require you to trot once round the ring but hunter judges only require you to go the length of the ring. Just before pulling up give the horse the command to slow down as, if obeyed, this will obviate the necessity of hauling on his mouth.

After your show, return to your place in the line and be ready to stand your horse up again should the judge come back down the line examining the horses. Most judges at this stage ask the whole class to walk round the ring again in a somewhat reduced circle, they then start to call the exhibitors into their final order of judging and then the rosettes are handed out.

It is only polite to thank the judge as he hands you the rosette and card, even if you are disappointed or have only received a commended.

This description of the method of showing horses in hand applies directly to youngstock and not specifically to mares and foals.

In the case of mares and foals the ring procedure is identical to the foregoing paragraphs except that, in all cases the mare and foal are both shown in the ring at the same time and must be led round the ring by two people, one leading the mare and the other the foal. The mare is usually led in front of the foal and judged first. At some shows the foals are judged immediately after the mares, while at others all the mare classes are judged first and the foals from all the classes are divided into their sexes and each sex (i.e. colts or fillies) is judged separately.

When you are called into the middle of the ring, whichever animal is being judged must be made to stand in front of the one which isn't being judged and when they are trotted out they should be trotted side by side and not behind one another, otherwise the judge will be unable to see the action clearly, of the one he is trying to judge.

When you have finished being judged and before returning to your loose-box make quite sure you are not wanted back in the ring to compete for a championship or special prize – all first and second prize winners from every class, except for foals are usually required to compete for the breed championship; if there is more than one foal class, foals are often required to compete for their own championship.

All prize winners are bound to parade at every show where a grand parade of prize winners is staged – failure to parade usually means the withdrawal of any prize money won – a steward is posted on the collecting ring or main ring gate to take down the numbers of all exhibitors who parade. This grand parade is usually held in the late afternoon of the day of judging or at some shows, on the next day, which means if you have won a monetary prize you must stay at the show for an extra night, in which case there is usually no extra charge for stabling.

It is customary not to plait up for a parade and indeed, most people remove all their plaits before the parade so that they can load up and go home immediately afterwards. Non prize winners are usually at liberty to go home straight after the judging, although some shows keep their gates shut until about 4 p.m. to enable the general public to see all the horses.

If you have won a major prize, after the judging a steward may come round to enquire something about your horse's past show record, this information is for the commentator during the grand parade and will be read out as your horse enters the ring.

When you are getting ready for the parade, as far as possible fix your horse's rosettes to the near side of the bridle, so that the crowd can see them as you walk round, any ribbons which might flap into the horse's eye should be anchored underneath the bridle, so that they don't frighten the horse. All rosettes from every class, won at that particular show should be worn for the parade, otherwise during the judging only rosettes won in the particular breed or type class should be worn, regardless of whether you are still showing under the same judge.

About half an hour before the time to parade you will be called down to a collecting ring, where you will be sorted out into class order: the breed or type champion and reserve champion from whatever class they came, usually lead the parade of that breed or type followed by the prize winners in order starting with the first prize winner of each class.

All the prize winners, usually walk once round the main ring and then line up across the middle of the ring, the cups and trophies are then awarded and the winners go forward to receive them. After the presentation of the trophies, the prize winners then file once more round the ring and out.

Small shows and gymkhanas seldom have veterinary surgeons on the gate, to check the health of the horses coming onto the showground. However, they do have an official on the gate to collect entrance money from the general public.

Your horse-box should be taken to the parking space allotted to these vehicles. This type of show, seldom if ever, provides stabling for its exhibitors, therefore the horses must remain inside the horse-boxes when not actually being shown, or got ready for the ring.

Quiet horses can be tied to the side of the horse-box with a hay-net, to keep them occupied, but this is not to be recommended for youngstock.

On arrival at the show, it is advisable to go to the Secretary's tent and make your entry, if you have not already done so and collect your ring numbers.

For one day shows it is absolutely essential to arrive at least an hour before you are due in the ring. If your horse will stand quietly while you plait it up at the show, it is a good idea to leave the tail until you arrive on the showground, as tails can easily be rubbed out while the horse is travelling. The mane is probably better put in the night before the show or early in the morning before you leave home. There are two disadvantages to plaiting manes up, the night before the show: horses tend to rub their manes and necks when they are left plaited up overnight, so that, even if the plaits don't actually come undone, they will look rather wispy by the morning; small pieces of straw tend to get entangled with the plaits, when the horse lies down during the night if it is kept in a loose-box for the night and these have to be picked out carefully in the morning. However, if plaiting up is left until arrival at the showground, it may mean a very rushed job, as there can be traffic hold ups in the immediate vicinity of the showground, on the day of the show.

At small shows it is unwise to rely on an official to tell you when to come into the ring for judging, so a very close watch must be kept on the progress of the ring events, otherwise you might miss your class. Some small shows do not have a collecting ring and the entries file straight into the judging ring. Many small shows have separate showing rings for in-hand and ridden classes, the ridden classes and show jumping always takes place in the main ring, whereas the in-hand classes often take place in a subsidiary ring.

The turn out and method of judging and showing at small shows, is identical to that already described for larger shows, only that small shows don't usually organise a grand parade later in the day, so after the judging is over the exhibitors are usually free to go home.

As far as many of the larger shows are concerned, where the loose-box charges are not included in the entry fees, horses may be brought to the showground on the day of the show – the latest time of arrival is usually given in the schedule. In this case the advice given for plaiting up and preparation for the small shows, applies equally here.

PART THREE

THE SALE OF BROOD MARES AND YOUNGSTOCK

There are two ways of disposing of animals you have for sale. The one is to sell them privately the other is to take them to a horse sale.

Selling privately is to be recommended wherever possible if you wish to have some control over who buys the horse. To sell privately an advertisement in the local press and possibly also in a horse magazine or paper, will often have the desired results. Alternatively, if the horse or pony in question, is a good class show animal, there are often several people only too willing to buy it after a good win in the show ring. An old horse dealer once told me 'never turn down a good offer for a horse, as if you do something is bound to happen to that horse later on'. Some breed society shows advertise horses which are for sale by organising parades for them at their annual show or putting the words FOR SALE under the horse's name in the catalogue.

Generally speaking preparation of youngstock for the sale ring is identical to the preparation necessary for the show ring.

Most youngstock look better if they are shown or offered for sale slightly on the fat side. A good layer of fat tends to cover up many conformation faults, as it rounds off the contours and makes the majority of horses look a better shape, except perhaps for rather cobby ponies, which if shown too fat tend to look rather stuffy.

Thoroughbred youngstock, bred for flat racing are usually broken at two years and therefore only appear in the sale ring as unbroken stock up to this age. Horses bred specifically for National Hunt racing are not broken until they are at least three years old and sometimes much later.

All foals and yearlings whether for sale or not should be fed as well as possible to obtain maximum growth and condition. Thoroughbred yearlings by tradition are brought to the sales fat but as far as the majority of trainers are concerned, a horse in good condition but not carrying any surplus fat is the better proposition. Fat layed down externally is also carried internally round the heart, lungs, etc., and must therefore be removed before the horse can be put into hard work.

Most horse sales are advertised in *Horse and Hound* throughout the year and in almost all cases a catalogue of lots to be sold is printed by the auctioneers concerned, some of the smaller sales accept entries on the day of the sale and also produce a catalogue. It is in the owner's interest to enter a horse early enough for the description to appear in the catalogue and so be read by everyone interested in the sale. Many people looking for a particular type of horse or pony, write for sales catalogues and go through them before making up their minds which sale to attend.

As far as unbroken youngstock are concerned, these are usually sold without any warranty for soundness.

Most youngstock with the possible exception of National Hunt stores – which should be sold, mouthed only; sell better as three year olds if they are mouthed and backed but unschooled, small children's ponies sell better if they are also schooled and can be ridden by a small child. Ponies under three years old should never be broken and ridden as even if they look adult they will lack maturity and with work could go lame.

Horses which are broken and known to be sound should have a recent vet's certificate or sold 'sound in wind, eyes, heart and action' or 'sound but subject to re-examination.'

For pure-bred animals a three generation pedigree should be given, supported by family details of those three generations and also mentioning any ancestors of interest further back in the pedigree. This is of particular importance in the case of the Thoroughbred racehorse which is often bought more on his racehorse potential than on his conformation alone, although the latter is of course of great importance.

These catalogue details are sometimes supplied by the auctioneers when you enter your animal or in the case of Thoroughbreds they can be obtained, for a small fee, from the majority of Bloodstock agencies. The seller (vendor) is ultimately responsible for the accuracy of the description and pedigree of their horse, which is now of major importance since the Misrepresentation of Goods for Sale Act '68.

Any horse entered for a sale which is primarily for Thoroughbred horses and which is not entered in the General Stud Book must be so described.

All animals up to four years old must be described as colts, fillies or gelding according to their sex, and after this age as stallions, mares or geldings. The only exception to this rule is where a filly, prior to four years old is covered by a stallion, in which case she must be described as a mare, even if she is only a yearling or a two year old!

In the case of a brood mare the exact date of her last service must be given and where the mare is sold as believed to be in-foal (mares are never sold guaranteed in-foal) this statement must be supported by a recent pregnancy certificate which should be given to the auctioneers before the sale.

Any horse which wind-sucks (the vice), crib-bites or has been operated on for unsoundness of wind must be declared as such.

If the buyer of a horse which has been sold with any kind of warranty or description contends that the horse in question does not correspond to that warranty or description he can return the horse to the auctioneers, usually within 48 hours of the sale, but in some cases as long as 7 days after the sale. The auctioneers or someone appointed by them will settle the dispute.

The horse ceases to be the property of the vendor from the moment the hammer drops and from that time becomes the sole responsibility of the buyer.

Most auctioneers charge an entry fee and take the guineas from the sale price – all bidding at horse sales being conducted in guineas and not in pounds, or they charge the guineas from the reserve price, if one is put on and the horse leaves the ring unsold.

The reserve price is the sum at which you are prepared to sell your animal. If you wish to have a reserve price on your horse you must declare this to the auctioneers before the time of the sale otherwise your lot will be sold without reserve to the highest bidder. Except for smaller sales where the reserve can be given to the auctioneer when the horse enters the ring. The reserve is never disclosed to the public, but the auctioneer will often take false bids up to the reserve figure in the hope that someone will come in and buy the horse.

At the majority of sales once the horse has been entered for the sale it must not be sold privately within 42 days prior to the sale or within seven days after the sale, unless a fee is paid to the auctioneers.

When preparing riding ponies for the sale-ring the main aim and object in view, is to make them look as much like a Thoroughbred as possible. There is a real art in producing horses for the show ring and sale yard but any person can improve the appearance of their horse greatly by a little careful trimming.

A Thoroughbred has very little, if any, hair around his heels, hooves, head or ears – the more highly bred he is, the less hair he has. The hair on his body is short and fine his mane and tail are usually thin, straight and silky.

The common bred horse is the absolute opposite to the above: his mane and tail are bushy, coarse and thick therefore they should be pulled and thinned, the mane can be plaited up on the day of the sale so it need not be made to lie on the right side of the neck – although it must be pulled on this side. He will probably have a beard and whiskers, any long coarse hairs round his eyes and on his muzzle can be cut off, as close to the skin as possible with a sharp pair of scissors and any fluff can be trimmed out of his ears in the same manner – but be careful not to let any hair fall into his ears, as this would probably set up an irritation which would cause him to shake his head. If there is only a small beard under his chin, this can be removed by singeing, but a thick beard and any hair in the heels should be removed by means of clippers – it must never be cut with scissors, as this leaves ridges.

To trim any surplus hair with clippers, insert a coarse blade into the clipper head, screw down the head but not so that it is tight, apply plenty of oil to the head and along the edge of the blades. Start the clippers working and allow the horse time to get used to the noise before you start trimming. When trimming run the clippers down the coat, in the direction of the hair – as opposed to clipping a horse – when you always run against the coat. This takes the long hair off, without making the horse look obviously clipped. Any hair round the top of the hooves can be removed either with the clippers or with a sharp pair of scissors.

Trimming should be done at least a week to ten days before a horse is going to appear in public, to allow the hair to grow a little and so look more natural, this is particularly important where white markings are concerned, as these are usually stained slightly yellow on the surface only and therefore show pure white in patches, where the hair has been trimmed.

Long coarse winter coats can be made to look better if they are trace clipped, but where it is known that the animal will be kept inside and rugged up, a hunter clip may be given leaving only the saddle patch and the legs. A horse or pony which grows a coarse summer coat can be made to look more quality if it is kept rugged up all the summer and/or if the coat is clipped in the early spring just as the new summer coat is coming through, this is an exception to the general rule and if practised on all horses, would ruin the summer coat of more finely bred individuals.

Breeds of horses and ponies which are normally shown unplaited and virtually untrimmed, i.e. in their natural beauty, and this includes Palominos, should be taken to a sale in the same way, in the case of this type of horse the tail should not be cut level at the bottom but if it gets too long it should be pulled shorter. Other horses should be plaited up and their tails pulled or plaited and cut level to the bottom of the hock joint for horses, or above the fetlock for ponies. The tail should be cut so that it slopes slightly upwards towards the horse's hocks and not straight across, in this way it will appear level when the tail hangs naturally. Thoroughbred horses entered for sales exclusively for this breed, except for those actually in, or just out of training, are always sold unplaited, but they should have their manes neatly pulled and layed, at ordinary horse sales and fairs they should be sold plaited up.

The tack used for sales should be the same as for the show ring, with the same standard of turnout for both horse and the attendant or rider, as in the show ring.

One of the most important things to remember about horse sales is to arrive early, so that prospective buyers have plenty of time to examine your horse before it goes into the sale ring. Most of the larger sales have stabling available for the horses and it is very wise to plan to make use of it, as in this case, some buyers come round the sale yard the night before, to view the lots.

At the larger sales it is customary to give additional information to that already printed in the catalogue and also the reserve price, height, registration, recent pregnancy and covering certificates to the auctioneer before the sale starts and then not to go near him again until after the sale. At smaller sales and horse fairs it is customary to stand up by the auctioneer's side while he is selling your animal and give him any information concerning it together with the reserve price. At these sales you can then agree to accept a lower price if the animal does not get to the original reserve, before it leaves the ring.

At all sales each horse is issued with a lot number, usually on payment of the entrance fee, at smaller sales these are usually printed on small oval paper discs which are stuck onto the horse's rump – one either side, but at the larger sales they are usually tied to the bridle each side – in which case make sure they are on the right way up as nothing looks worse than a number tied on upside down, at these sales the small numbers are only stuck onto the horse's rump after it has been through the sale ring.

At the smaller sales, the horses usually come into the ring on the auctioneer's right and stand in front of him while their description is read out : this usually includes their age, sex, height and whether they are broken or unbroken (at some sales unbroken stock are sold in a separate ring), if they are broken – whether they are quiet to ride and/or drive and any prizes or competitions they may have won.

The animal is then turned round and trotted out smartly to the end of the ring, turned and trotted back. This is usually repeated at the request of the auctioneer if there is a pause in the bidding. The usual dealers' term for trotting a horse out in-hand is 'give us a yard'.

At the larger sales, the animals are walked round in a circle before they actually go up into the sale ring, here they can be pulled out and inspected by anyone who wishes to do so. At all times these horses must therefore be walked out smartly and although it is not a customary practice in the Thoroughbred world, a short stick, as carried in the show ring, can be a decided advantage in obtaining an energetic walk with the horse's head up. In the sale ring the horse must be walked round very smartly on the right rein – it will not be required to trot, although a slow jog is better than a slow shuffle but a lively walk is ideal.

All lots must be payed for before they leave the sale yard. Payment to the vendor for lots sold is usually made by the auctioneers automatically a few days after the sale but some firms require a written request for payment from the seller or his representative, before any money changes hands. It is therefore just as well to read the sales conditions in the catalogue.

11

SOME DISEASE CONDITIONS OF YOUNG FOALS

Unlike the human baby the foal does not receive any resistance to disease before birth and has to acquire all its immunity from the colostrum in the first 24 hours after birth, the colostrum or fore milk contains high levels of antibody and the ability to absorb these across the small bowel ceases before 36 hours of age. Antibodies are substances found in the blood, which destroy germs that would otherwise cause disease. After 36 hours the foal can only obtain immunity artificially by means of injections or by natural development over a period of time.

When dealing with the problem of disease, prevention is always better than cure and is often far cheaper in the long run, so wherever possible I will point out the preventive measures available to the owner for all the more common conditions and diseases met with in young foals as well as the symptoms and when to call in your veterinary surgeon.

BARKING and WANDERING

This is an obscure condition in new-born foals, in which the foal appears perfectly normal at birth but sometime afterwards, in some cases it will start to make a noise like a small dog barking and will show marked signs of respiratory distress. If the foal is already lying down he may jerk his head up and down and move his legs without being able to get up; if he is already on his feet he will probably show definite symptoms of blindness: he will wander around aimlessly, bumping into the walls and continually calling for his mother without apparently being able to recognise her.

The foal should be placed on a blanket under a lamp for warmth if possible and veterinary aid must be sought immediately before there is permanent brain damage.

The exact cause of this condition is obscure, it may be attributed to several factors such as infection by the germ Actinobacillus equuili, or other organisms such as those which can cause septicaemia. Lack of oxygen at the time of birth associated with premature breaking of the cord or fractured ribs sustained at the time of birth, could also possibly cause this condition.

This is a very rare condition except in Thoroughbreds.

ENTROPION

This is a condition in which the foal exhibits a chronic watering of one or both eyes. On careful examination it will be seen that the bottom eye lid has turned inwards and the lashes are irritating the eye.

This is a job for your veterinary surgeon and he may wish to insert some stitches into the bottom lid which will make the lid turn outwards in the normal manner, these are left in place for about two weeks.

HAEMOLYTIC DISEASE

This is a blood disease of the newly-born foal, in which the chief symptoms are anaemia and jaundice. The disease is caused by a blood group incompatibility between the foal and dam and is analagous to Rhesus babies as seen in human medicine.

The condition arises when a foal, while still in the womb, inherits a blood group from its sire which is incompatible with the dam's blood. The mare may then become sensitised to this blood group and produce antibodies against it. She concentrates these antibodies in her colostrum and on suckling they pass into the foal's blood stream; the antibodies then combine with and destroy the foal's red blood cells, which results in severe anaemia and jaundice.

If you own a mare which has had a foal with haemolytic disease or if you want to make sure that your mare does not have a foal with haemolytic disease, then about three to four weeks before she is due to foal, ask your veterinary surgeon to take a sample of her blood. He will send this away to a laboratory to be tested against samples taken during the succeeding two weeks, for an increase in antibody level. Any significant increase in level would denote that the foal will be at risk should it receive any colostrum from its own dam.

If it has been determined that the mare has an increased antibody level and that the foal when it arrives will therefore be at risk, the main preventative measures to take are :

(i) Obtain a foal size muzzle – this must be strapped on to the foal's head the moment he gets to his feet, to prevent him from drinking any of his mother's milk and must be left on for 36 hours. If all access to colostrum is withdrawn the disease cannot occur.

(ii) As the foal must have some colostrum an alternative supply should be provided, this can be obtained at any time from another unsensitised foaling mare and kept in a deep freeze until ready for use, it should be warmed to blood heat before feeding. Some veterinary surgeons in equine practise keep a colostrum bank for emergency use.

The mare must of course be milked out at least six times a day to prevent her from drying up before the foal is allowed to suck. This milk must be discarded.

The clinical signs of this condition are that the foal will be seen to yawn, appears sleepy and short of breath due to lack of oxygen in the blood, from the destruction of the red blood corpuscles. The mucous membranes are yellow. This condition usually becomes apparent within the first 36 hours after birth.

The foal unlike the human baby receives no immunity while in the womb, so if the foal does not receive any colostrum from its own dam NO haemolytic diseases can occur. By examining in late pregnancy for antibody, a foal need never suffer from this disease.

HERNIA

A hernia or rupture is caused by incomplete development of a muscle wall (usually abdominal) and the protrusion of an

Hernia

organ or it's fatty connective tissue through the hole which has been formed, causing a swelling to the outside beneath the skin.

The type of hernia most commonly met in foals is the umbilical hernia. This is seen as a swelling in the naval region and is due to part of the bowel protruding through the umbilical ring which has failed to close up in the usual way.

This type of hernia should not be confused with thickening which sometimes takes place in the region of the navel and which is hard and unyielding, whereas in the case of a hernia a distinct ring can be felt through which the protruding mass can be pushed back into the abdomen and through which it protrudes again once the pressure is released.

An umbilical hernia is sometimes self curative and tends to disappear quite naturally as the animal gets older, but in some more severe cases surgical treatment to close the hernial ring is necessary.

Some people advocate applying blisters to the area with some success but this tends to thicken the skin making surgery if necessary more difficult.

MENINGITIS

This is the word used to describe inflammation of the membranes covering the surface of the brain, and/or spinal cord. It can be a sequel to severe infections of the upper respiratory tract, e.g. Strangles.

The affected animal usually becomes dazed and is liable to walk into objects in its path. It appears ill – the temperature, respiration and pulse are higher than normal. Veterinary help should be called in immediately as this is a very serious disease and only very careful nursing can effect a cure.

NAVEL OR JOINT ILL

This is a disease characterised by swelling around the region of the navel and/or swellings in the leg joints accompanied by marked lameness in the affected limb or limbs.

The germs which cause this disease are thought to gain entry at the time of or soon after birth through the navel cord, or even from the mare before birth. There is particular danger when the cord is broken prematurely, as the umbilicus does not have time to close up immediately as it does when the cord breaks naturally. For this reason, as well as for reasons of loss of blood, the cord should be left to break naturally and maximum cleanliness should be observed at the time of foaling to cut down the possible risk of Joint Ill occurring, from this point of view it is far better to foal a mare outside in a field than in a dirty loose-box.

The symptoms of Joint Ill are usually that the foal goes off suck and appears dull and listless, the temperature rises to 103–105°F. and the foal will tend to spend most of its time lying stretched out on its side.

The navel may be wet and oozing a blood-stained fluid or it may be dry, swollen and obviously painful, due to the formation of abscesses.

Some of the foal's joints may also be swollen, tense and hot, if left untreated these will eventually burst and discharge blood-stained material, the foal will gradually get weaker and die, in some cases the abscesses are also found on the internal organs in which case death is very rapid.

Prevention mainly consists of maintaining scrupulously clean surroundings at foaling time and a liberal application of sulphonamide powder to the cord the moment it breaks.

Should you suspect that your foal might have Joint Ill, your veterinary surgeon should be called in immediately. No other mare should be allowed to foal down in the same loose-box as one that contained a foal with Joint Ill unless it has been thoroughly disinfected first, as this is a highly contagious disease – and very difficult to cure unless treated at once.

PERVIOUS URACHUS

In this condition a continuous trickle of urine from the navel will be observed, which will probably take the hair off the lower regions of the hind legs.

Before birth the foal lies inside the sac known as the Amnion and is immersed in Amniotic fluid, this is surrounded by the chorio-allantois membrane which contains allantoic fluid, this is the foetal urine which has collected before birth, the urinary bladder opens directly into chorio-allantois which thus prevents over-distension of the bladder before birth, at the time of birth, however, the umbilicus closes.

In the case of pervious urachus the urachus continues to have direct access via the navel cord with the outside and does not close up in the usual way.

This condition is sometimes self curative but if it does not seem to improve within two weeks your veterinary surgeon should be consulted as treatment may be necessary. Vaseline should be applied to the hind legs to prevent blisters forming from the action of the urine on the skin.

PNEUMONIA

This is inflammation of the lungs with a resulting breakdown of the lung tissue which greatly impairs respiration. A variety of germs: bacteria and viruses, are able to cause pneumonia in susceptible young foals.

Predisposing causes of pneumonia may be exposure to cold, wet weather, or from stuffy and badly ventilated stables. It can also be produced by careless drenching – allowing some of the medicant fluid to go down onto the lungs.

The possible symptoms of this disease are: a high temperature, fast laboured respiration, the foal will go off suck and there will probably be fits of shivering, there may be a nasal discharge. In the case of pneumonia the care taken in nursing and management is of utmost importance as a relapse can take place at any time if the standard of care is relaxed.

A veterinary surgeon must be called in at the very beginning if a good chance of recovery is to be expected.

RETAINED MECONIUM COLIC

Meconium is the brown or black hard excretory material which collects in the bowels before birth. It should be passed within a few hours after birth. The passing of the meconium is accelerated by the purgative action of the colostrum. The meconium is thought to consist of bile and debris from the intestines which has collected before birth.

It is very important that the meconium should be passed

within a few hours of birth to allow normal digestion to take place. If the meconium is retained, the foal will probably carry its tail higher than normal, it may be seen standing in a crouched position with its back arched straining; in a more advanced stage it will show very definite signs of colic: i.e. kicking up at its stomach, rolling and even lying on its back with its legs in the air in an effort to relieve the pain.

Retained meconium and constipation in the foal is often associated with a costive diet in the mare, therefore as far as possible the mare should receive some green stuff, carrots or mashes every day as well as adequate exercise.

In the early stages of retained meconium colic the foal can be given a dose of liquid paraffin. It is probably a better plan to call in your veterinary surgeon immediately you see the first symptoms.

An enema of warm soapy water does sometimes help to relieve this condition.

SCOURING

This is another word for the condition known as diarrhoea, a symptom of several diseases or errors in feeding.

The foal will often scour when the mare comes in-season due to some change in the composition of her milk at this time, which has an irritating effect on the bowel, this usually stops once the mare goes out of season. In this case the foal does not go off suck. A useful tip is to smear vaseline round the foal's hindquarters to prevent the hair from coming out, this is especially important if you intend to show the foal.

Scouring at other times in young foals should always be treated immediately by a veterinary surgeon, as young foals have very little reserves and are soon pulled down in condition. Some cases of scouring are due to bacteria which can be identified by taking a swab from the rectum and growing the organisms in the laboratory. Any foal which is seen to be scouring should be isolated immediately as this disease can spread to other foals very rapidly.

Draughts and sudden changes in temperature can be predisposing causes of scouring. Other types of scouring can be due to digestive upsets.

Of major importance in all cases of scouring is an adequate intake of fluid to replace the loss to the bowel, at the same time certain vital substances known as electrolytes which are lost during scouring must be replaced. This is often done by the veterinary surgeon by means of a stomach tube. If the fluids are not replaced in adequate amounts the foal will die as a direct result of fluid loss. The foal's tongue usually becomes coated and the eyes may sink into their sockets, there is a marked loss in condition and wasting of the muscles.

Treatment usually consists of the administration of antibiotics by injection or by mouth, and foals usually recover without any permanent ill effects. Injection of some of the mare's blood into the foal will sometimes give an added boost to recovery due to the antibodies this blood will contain.

SEPTICAEMIA

In this disease bacteria have somehow managed to gain entry into the blood stream, where they circulate round the body and invade any or all of the vital organs. A predisposing cause of this disease can be lack of colostrum and thus the antibodies it contains. In most cases the animal dies, the course of the disease usually being very short.

Sometimes the only symptoms noticed are a high temperature and general signs of illness followed soon afterwards by death. The sooner a veterinary surgeon can be called in the more chance he will have of saving the animal's life.

STRANGLES

This is a disease more commonly seen in young horses up to five years old. It is caused by the germ, *Streptococcus Equi*, and is very highly contagious; incubation is usually from two to fourteen days. Infected horses often show a rise in temperature, appear dull and go off their feed. They often have a cough, sore throat and profuse discharge from the nostrils and swellings in the angle of the jaw. This causes the horse to hold its head and neck out stiffly. After a few days an abscess will form in the swelling, mature, and finally burst, discharging bloodstained pus. This pus and the nasal discharge is highly infectious, so very great care should be taken that anyone handling an infected horse does not come in contact with other horses at this time. As soon as the abscess has burst the temperature usually falls and the swelling subsides. Recovery is usually complete in a few weeks.

Treatment consists in immediate isolation of the infected animal and calling in your veterinary surgeon, who may wish to treat the animal with antibiotics.

TETANUS (OR LOCKJAW)

This disease affects horses of all ages but its incidence among young foals can be reduced to a very large extent by giving a mare which has already had primary Tetanus toxoid injections a booster injection about 3 weeks before foaling, she will then produce antibodies, which will be transferred to the foal in the colostrum, giving the foal immunity to tetanus for the first few weeks of its life.

This disease is caused by a germ which can survive for long periods outside the human or animal body. It does not utilise atmospheric oxygen in order to live, this germ gains entry mainly through deep wounds, such as those in the feet, any deep puncture wound or through the navel at the time of birth, (i.e. any site not immediately in contact with the atmosphere).

The germs do not leave the site of entry but produce a toxin which is liberated into the blood stream this acts on the central nervous system producing symptoms of paralysis:

The horse exhibits unusual signs of nervousness.

He will stand with his limbs stiff, tail raised stiffly, head and neck stretched out and ears pricked.

If the head is raised quickly the third eyelid (haw) will flick across the eye to a far greater degree than is usual.

If tetanus is suspected your veterinary surgeon must be called immediately. Wherever possible, preventive measures should be taken. These fall into two groups:

i.e. The injection of Tetanus serum and Tetanus Toxoid.

Tetanus serum is the faster acting of the two but only provides immunity for about four weeks, it has fewer side effects than the toxoid and is consequently invaluable for administration to young foals out of mares which have not themselves been injected with Tetanus toxoid, it is also used after accidents where animals have cut themselves and are not immune to Tetanus.

The toxoid is slower acting and is used to give lasting immunity, it is given at three months old with a second injection one month later, this is followed by a booster injection a year after the last injection to give life-long immunity but where an animal is at risk for instance in the hunting field, an annual booster injection is often a wiser precaution.

Young horses should always be injected against tetanus as they are far more prone to accidents and injuries than older horses.

TYMPANY OF THE GUTTURAL POUCH

This a very rare condition sometimes met in foals. The affected animal appears as in the diagram below with a large swelling in the region of the gutteral pouch. When pressure is applied to the swelling air is driven out but the swelling returns within a few hours.

The affected animal makes a roaring or snoring noise on breathing and will probably make more noise when eating. In hot weather the animal will be seen to find great difficulty in breathing. There may or may not be a nasal discharge. This condition may be hereditary or it may be due to an infection. Immediately you see the first symptoms, you should call in your veterinary surgeon and he may advise an operation. On no account should you attempt to lance the swelling yourself to let any fluid or air out, as this could start up a serious infection.

These days this condition is usually curable.

Tympany of the gutteral pouch

WEAK AND DEFORMED LEGS

Foals like babies can be born with various weaknesses and deformities of their limbs.

Some of those more commonly seen are :

(a) Over long weak pasterns, where the foal walks with his fetlocks touching the ground. This condition is usually self curative, as the foal becomes older and stronger the pasterns tend to strengthen. Reducing the length of the toes of the affected legs does seem to help.

(b) In the case of contracted tendons the foal may only be able to walk on the tips of his toes, in very bad cases the foal may be entirely unable to get up, if the contraction is very bad the foal may have to be put down, but in some cases an operation to cut the tendons and straighten the leg, may be successful, in other cases splinting the leg will often work well.

(c) When foals are born with twisted legs which do not start to straighten within a week of birth, or where their legs start to twist soon after birth, your veterinary surgeon should be called immediately as the longer these weaknesses are left the worse the condition will become.

Feet which turn in or out can be corrected by a good blacksmith to a large extent but where the whole leg turns a splint will probably be needed to correct the fault.

SUMMARY OF DISEASE CONDITIONS IN FOALS.

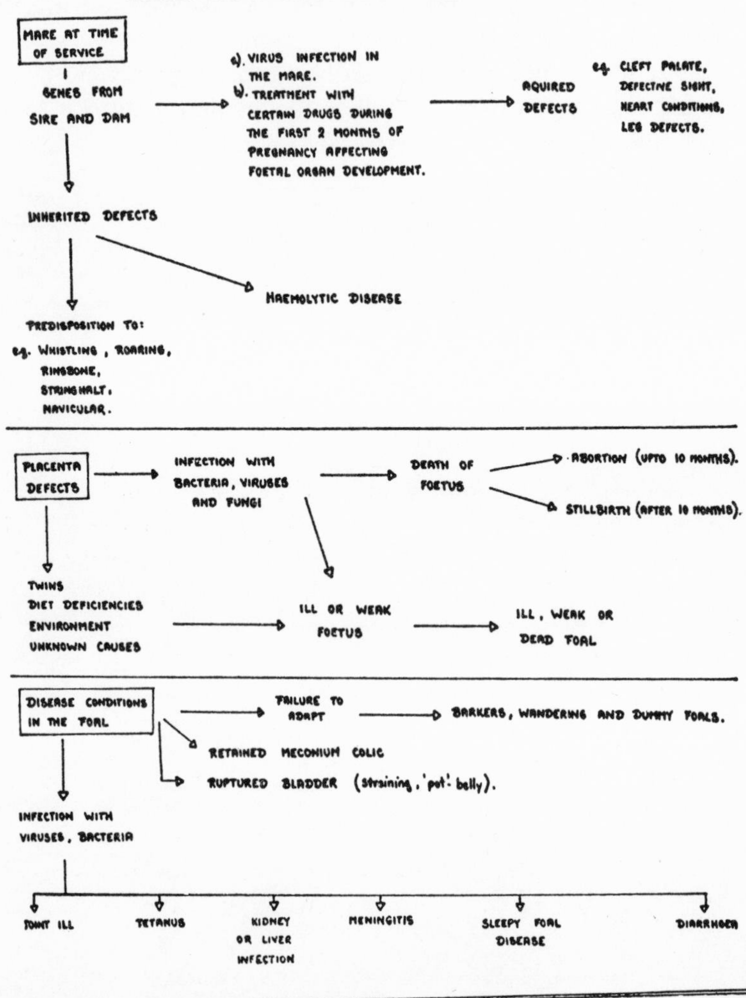

MARE AT TIME OF SERVICE
GENES FROM SIRE AND DAM
a). VIRUS INFECTION IN THE MARE.
b). TREATMENT WITH CERTAIN DRUGS DURING THE FIRST 2 MONTHS OF PREGNANCY AFFECTING FOETAL ORGAN DEVELOPMENT.
AQUIRED DEFECTS
e.g. CLEFT PALATE, DEFECTIVE SIGHT, HEART CONDITIONS, LEG DEFECTS.
INHERITED DEFECTS
HAEMOLYTIC DISEASE
PREDISPOSITION TO:
e.g. WHISTLING, ROARING, RINGBONE, STRINGHALT, NAVICULAR.
PLACENTA DEFECTS
INFECTION WITH BACTERIA, VIRUSES AND FUNGI
DEATH OF FOETUS
ABORTION (UPTO 10 MONTHS).
STILLBIRTH (AFTER 10 MONTHS).
TWINS
DIET DEFICIENCIES
ENVIRONMENT
UNKNOWN CAUSES
ILL OR WEAK FOETUS
ILL, WEAK OR DEAD FOAL
DISEASE CONDITIONS IN THE FOAL
FAILURE TO ADAPT
BARKERS, WANDERING AND DUMMY FOALS.
RETAINED MECONIUM COLIC
RUPTURED BLADDER (straining, 'pot' belly).
INFECTION WITH VIRUSES, BACTERIA
JOINT ILL
TETANUS
KIDNEY OR LIVER INFECTION
MENINGITIS
SLEEPY FOAL DISEASE
DIARRHOEA

12

CONDITIONS ASSOCIATED WITH FOALING AND LACTATION IN BROOD MARES

BRUISING AND LACERATION OF THE VAGINA AND VULVA:

This is a fairly common complication during foaling, and can occur during the birth of a large or abnormally presented foal, especially in very young mares or mares with a small pelvis. Laceration and tearing of the vulva will occur if a stitched mare is not cut enough to allow the foal to be born.

Damage by the foal as it leaves the birth canal will soon be noticeable: the lips of the vulva swell, appear congested and painful. The vaginal wall will also be reddened and somewhat inflamed. Examination of the mare in this case after foaling by your veterinary surgeon, is a wise precaution, to help prevent further trouble later on.

The risk of infection to the damaged tissue will be increased and this could lead to a metritis and cervicitis, possibly resulting in an infertile mare.

Mares which have sustained some bruising and lacerations at foaling should not be covered at the foaling heat, but should be left to some future date when they have healed up sufficiently. These mares often show signs of an acute and sometimes profuse vaginal discharge, with veterinary treatment this usually clears up quite soon and the mare will probably, then be ready to serve next time she comes in-season.

DAMAGE TO NERVES IN THE PELVIC REGION OR TO THE PELVIS ITSELF:

This causes a condition known as post-partum paralysis, which is probably more commonly seen in the cow than the mare, but nevertheless does occasionally occur in the mare.

Damage to the pelvis itself, is much more common than damage to the nerves and results from a bad foaling or slipping on a concrete floor.

In a severe case the animal is unable to rise after foaling due to the injury. The animal appears to be weak in its back or hind-quarters but is otherwise normal. In milder cases the mare can stand but is unsteady behind and unsure of herself.

Your veterinary surgeon should be consulted immediately the symptoms are observed as there may be a fracture of the pelvis, also rupture of the bladder can be a sequel to post-partum paralysis.

It is unwise to leave the foal in a loose-box unattended with a mare which is down, as, in her efforts to rise she might easily fall on the foal and injure it. The foal should be removed to another loose-box equipped with a heater and if possible a foster mother should be found, otherwise it must be hand reared. The mare can be milked out every two hours and the milk given to the foal, this *must* be done, at least until the foal has received colostrum for two days, then a mare's milk substitute may be given if preferred.

EQUINE INFLUENZA:

In these cases the affected horse will cough and run a temperature. The disease is only of consequence as far as ridden or working horses and young foals are concerned. Influenza can prove fatal in the case of very young foals, which under normal circumstances have very little resistance at this early age, unless

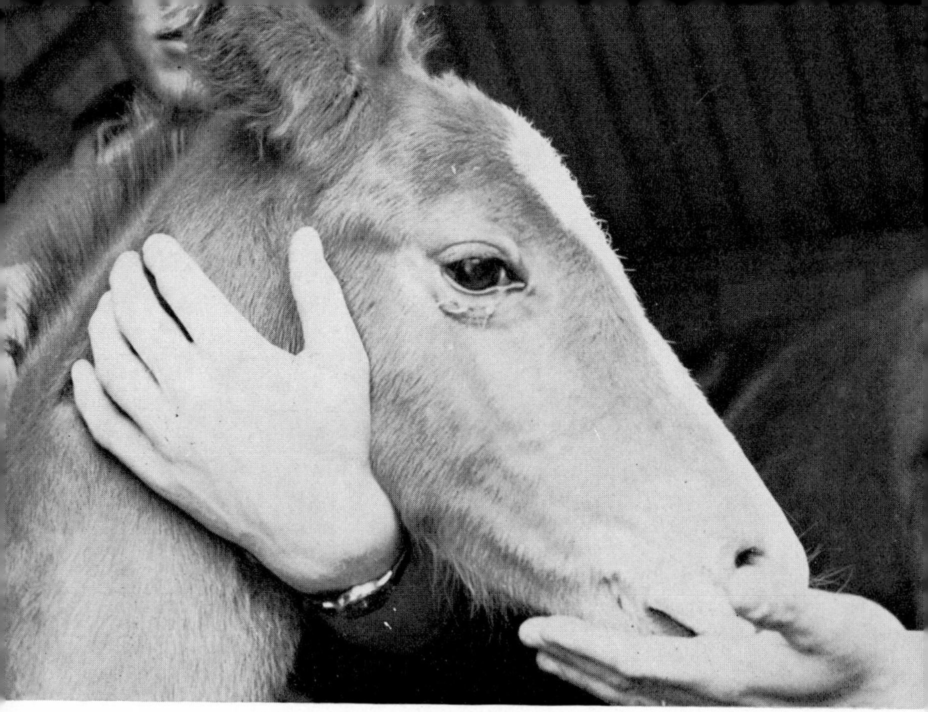

Above: A foal showing the effects of Entropion (*see* chapter 11, page 114)
Below: A foal with contracted tendons, showing maximum extension of the front legs (*see* chapter 11, page 125)

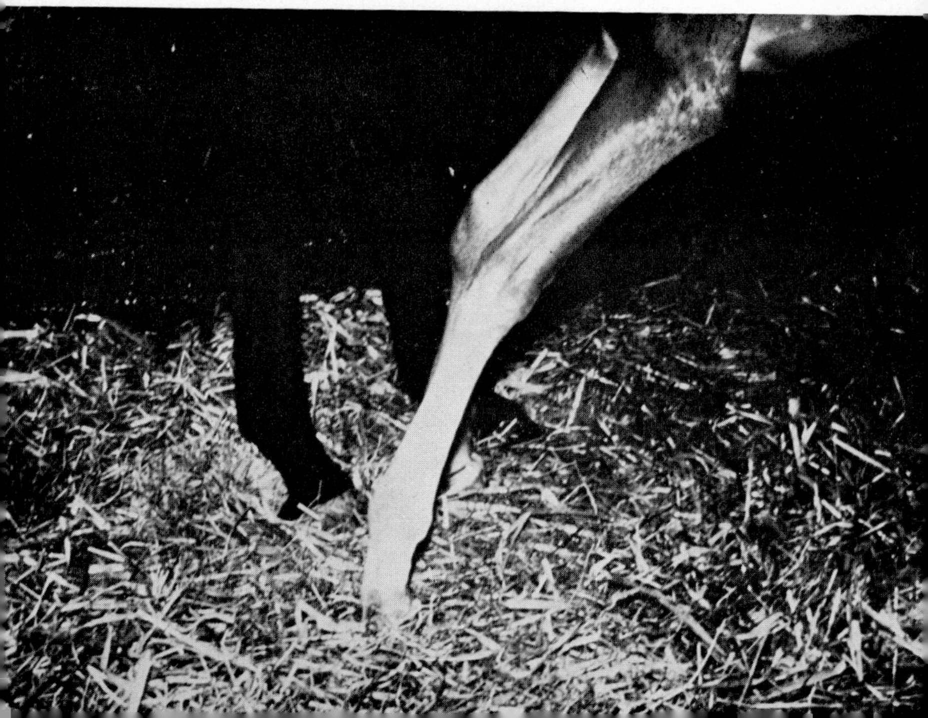

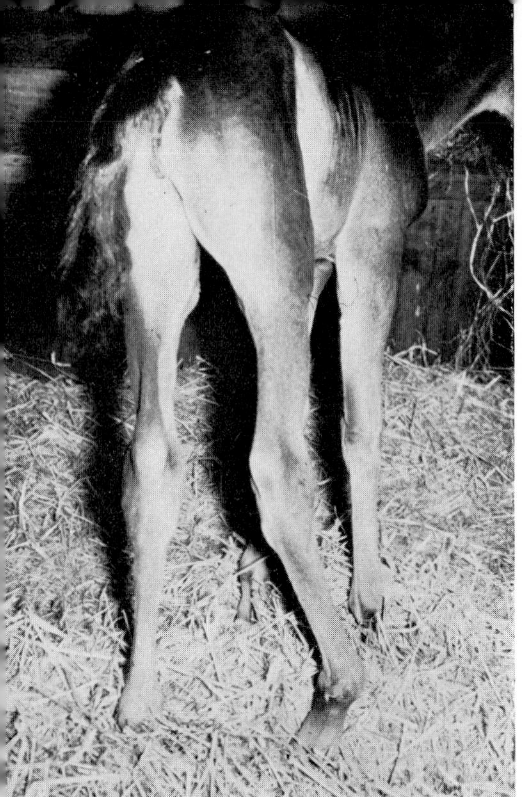

Another foal showing contracted tendons

A foal born with very weak front legs causing it to stand right back at the knee. Inset: the same foal two weeks later

their dam has been vaccinated recently. Its incidence among foals can be greatly reduced by making sure that all brood mares receive their primary injections against Equine Flu followed by a booster dose within the six month period before foaling. The mare will then produce antibodies, which are transferred to the foal in the colostrum giving it immunity to influenza for the first few weeks of life.

EQUINE VIRUS ABORTION:

This disease is endemic in some countries like America but sporadic outbreaks do occur in the British Isles from time to time. With the great increase in air travel for horses throughout the world the risk of an outbreak at anytime is very real. Vaccination of all foaling mares is routine practice on most well-run studs in the United States, but as a live vaccine is used for this purpose, vaccination is not permitted in the British Isles. This virus is identical with that causing 'snotty noses' in young horses – a condition usually occurring on studs each autumn and winter.

Abortions occur in infected mares between the fifth and eleventh months of gestation, most occurring during the ninth and tenth months. Infected foals carried to full term are usually born dying or in a very weak state.

The incubation period appears to be 20-30 days, which accounts for abortions occurring after 'snotty nose' attacks in young horses on studs from mid-October to March; young stock therefore appear to be the main cause of infection among foaling mares and wherever possible should be segregated.

Immunity to respiratory infection caused by this disease only lasts for three to five months, but immunity to abortion appears to last longer, as mares will often show evidence of respiratory disease when in contact with infected youngstock but will seldom abort as long as their level of natural immunity is maintained by regular exposure to infection. Abortion only occurs in stock whose level of immunity has been permitted to wane, thus allowing the virus access to the foetus.

Abortion does not appear to occur in mares infected with the virus in the early period of gestation. Where abortion does occur the mare does not make the usual preparation for foaling and seldom has any milk; she usually aborts quickly and easily. The afterbirth is seldom retained. The foetus usually appears very fresh and the 'bag' is seldom broken.

The foetus, fluid and membranes contain the virus and must therefore be destroyed completely or sent intact in a waterproof bag to a laboratory for testing. The virus usually dies in about two weeks, but on horse hair it can live for forty-nine days. In suspected cases, all the bedding etc. must be burnt and the loose-box thoroughly disinfected. The mare herself must be washed down with disinfectant; special attention should be paid to her hindquarters and tail.

The foaling heat should be missed but infected mares can be covered safely at the next heat, when they will often conceive and foal down normally the following year.

It is absolutely essential that any mares arriving from abroad including Ireland should be isolated until your veterinary surgeon is satisfied that there is no risk of infection.

HAEMORRHAGE OF THE UTERINE ARTERIES:

This condition usually occurs within a few hours of foaling. In the majority of cases there is no sign of any blood escaping from the vulva. When the mare is suffering from an internal haemorrhage and is losing a large quantity of blood, the mucous membranes (i.e. the gums and eyes) become pale and anaemic looking. The mare may sweat profusely and show signs of colicky pain. As she gets weaker, she will be unable to stand and may finally become unconscious.

In some cases the bleeding will stop automatically but in more severe cases of haemorrhage the outlook is very grave, however blood transfusions can sometimes be given, but in the case of severe haemorrhage these may probably be of no use.

This condition is more common in older mares, i.e. 10 years and over which have had several foals and more so those with very dipped backs and dropped bellies, and is probably due to the stretching of the uterine ligaments and arteries. The actual haemorrhage being caused by rupture of one of the arteries to the uterus due to damage in the wall of the vessel.

MASTITIS:

As compared with the cow this condition is comparatively rare in the mare, but nevertheless does occur occasionally.

Mastitis is inflammation of the udder tissue, due to a bacterial infection; this can be brought about by wounds or scratches on the teats and udder which become contaminated by flies or from lying on dirty bedding. It can also occur on occasion as a secondary infection to a disease such as strangles.

The mare's udder will appear hot, swollen and quite hard to the touch, due to the extreme pain in this region. She will probably not let her foal suckle and may attempt to bite or kick it should it go near her udder. Her temperature, respiration and pulse will be higher than normal, she may go off her food and be subject to occasional fits of shivering. If some milk is drawn from the udder, it will appear thin and watery and will contain several clots.

The mare will probably tend to walk stiffly with the hind leg on the affected side. In bad cases the swelling may extend forward from the udder, right along the entire length of the stomach to the front legs.

It is imperative that your veterinary surgeon should be called in, as soon as the first symptoms are noticed, as if neglected, this disease will lead to the complete destruction of udder function on the affected side.

Mastitis is very occasionally found in filly foals and maiden or barren mares but most often occurs at weaning time or when mares with foals at foot are first turned on to very rich pasture, which produces a flush of milk.

METRITIS:

This is inflammation of the uterus which may be either acute or chronic. In the chronic form the mare becomes permanently infertile, while in the more acute form the mare may recover with time.

When acute metritis occurs immediately after foaling it is usually due to the introduction of septic bacteria (germs), into the uterus during foaling, these organisms are usually conveyed by the unwashed hands and arms of the attendants or from unsterile ropes and instruments used to assist the foaling. It can also arise from a portion of retained afterbirth, this is usually a piece of the non-pregnant horn (see Chapter 5, p. 51), which remains attached inside the mare. Occasionally, due to the great weight of the portion lying outside the mare, the non-pregnant horn is torn through and is thus retained.

Metritis is a very severe condition in the mare and if left untreated is fatal. Your veterinary surgeon should be called in immediately the first symptoms are observed. The mare becomes very ill and loses all interest in her foal, usually within the first two days after foaling. There is usually a blood-stained grey discharge from the vulva, which will soil her tail. In very bad cases the mare may appear tucked up and stand around with her back arched, in obvious pain.

As far as metritis is concerned, prevention is better than cure, extreme cleanliness should be observed at foaling time. If it becomes necessary to insert your hands and arms into the vagina during foaling you must make sure that they are scrubbed clean in a disinfectant and your nails are short and free from dirt. All instruments and ropes used for foaling must be sterilised, preferably by boiling for 10 minutes, after which they should be left covered in the container in which they were boiled until ready for use. Loose-boxes used for foaling must be swept clean of dust and cobwebs and the straw used for foaling must be the cleanest available.

RECTO-VAGINAL FISTULA:

This condition is fortunately uncommon but may occur when due to abnormal presentation of the foal in the birth canal, one foot is pushed through the wall of the vagina and into the rectum. It is most common in first foaling mares.

The symptoms in individual mares will depend on the extent and position of the injury. In some cases the foot may be seen protruding through the rectum.

Where possible, the foal should be pushed back inside the mare. In any case your veterinary surgeon should be called in immediately and the mare got on her feet and kept walking round until he arrives, to prevent her from straining.

In most cases an operation to repair the tear can be performed with success.

RETENTION OF THE AFTERBIRTH:

In this case the afterbirth is not expelled in the normal way by the secondary birth pains, after foaling, but remains attached inside the mare. It must be removed within twelve hours of foaling or infection may set in, leading to a metritis and a possible general septicaemia with fatal results. In this respect the mare is far more susceptible than the cow, which can often be left for days without any apparent ill effects.

On no account should the owner attempt to remove the afterbirth himself, as traction on the visible part of the membranes would nearly certainly lead to some of the attached portions, tearing inside the mare. These portions would be retained and decompose which would lead to an acute infection and probable death of the mare.

As a general rule of thumb, if, when the mare has foaled during the night, the membranes are still retained first thing in the morning, your veterinary surgeon should be notified immediately, before he goes out on his rounds. When he arrives he will need a clean bucket of warm water, some soap and a clean towel.

UTERINE PROLAPSE:

This can occur anytime up to about 24 hours after foaling. In some cases only the vagina appears outside the walls of the vulva but in other cases the whole of the uterus invaginates and appears hanging downwards, as a large fleshy mass, often reaching as far as the hocks. This condition can follow an easy foaling and is probably more common in mares with high set tails and slack ligaments round the vulva.

A partial prolapse of the uterus can occur internally when one of the horns turns in on itself, in this case the mare will show very definite signs of colic, she will sweat profusely and get down and roll in an effort to ease the pain. There will be no other visible sign that there is anything wrong. Your veterinary surgeon should, however, be called in immediately.

If the whole of the uterus is prolapsed, it must be kept clean and warm until the veterinary surgeon arrives, and great care should be taken to prevent its getting damaged in any way. The easiest way to do this is to place the mass on a clean sheet and remove all the particles of straw which may be adhering to it. Clean warm water may be sprinkled over the uterus to prevent it from drying up while you await the arrival of the veterinary surgeon, as the longer it remains outside the body the more swollen and dry it becomes.

Mares have recovered after a uterine prolapse but generally speaking unless treatment is prompt the outlook is grave.

UTERINE RUPTURE:

This is a very serious although fortunately rare condition which can occur at any time during pregnancy but most especially at foaling time. It is most probably associated with a severe kick, blow or more commonly with a bad fall. It can occur during foaling due to the foal putting one of its feet through the wall of the uterus.

Rupture of the uterus invariably means that the animal will die unless an abdominal operation can be performed to close the wound, often a very difficult procedure.

SUMMARY OF DISEASE CONDITIONS ASSOCIATED WITH FOALING AND LACTATION.

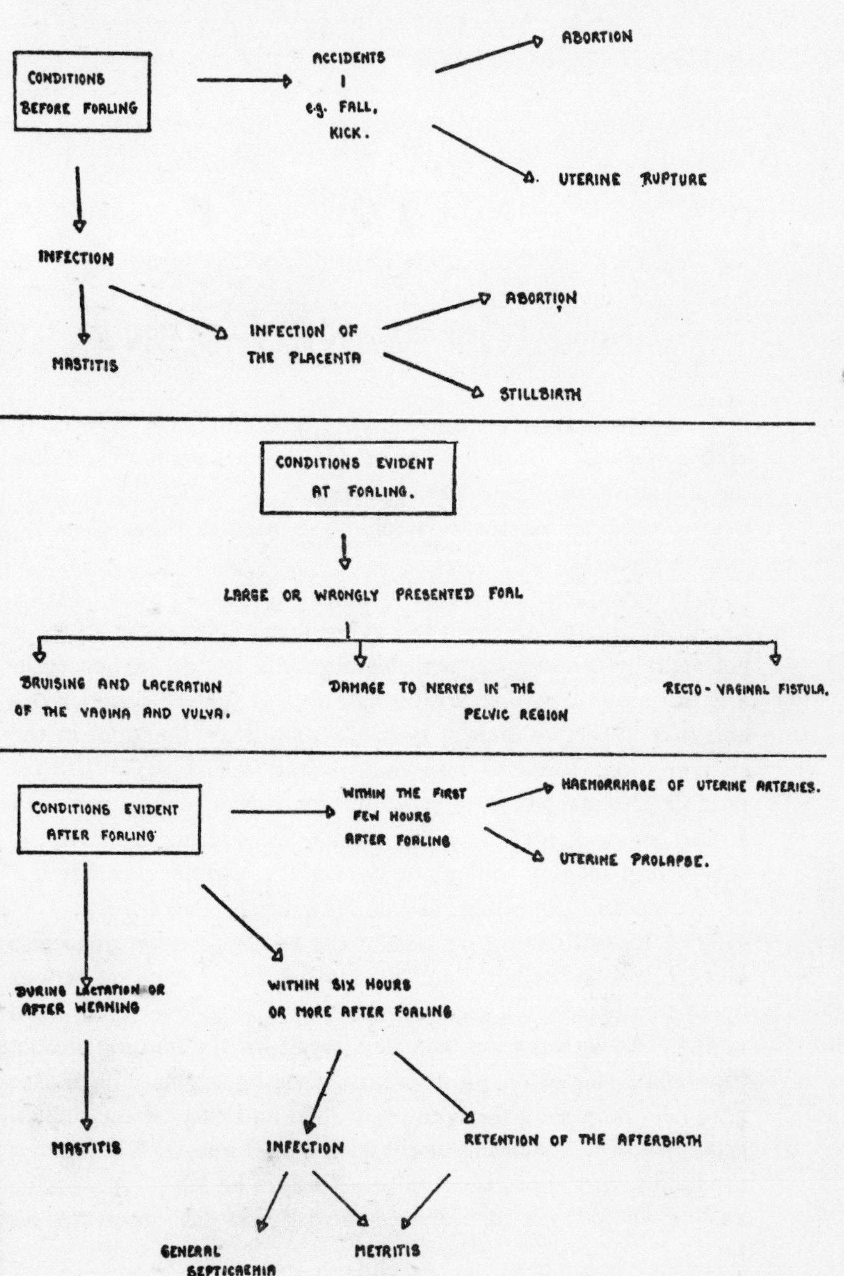

13

STALLION MANAGEMENT

It would be false economics for the amateur horse breeder to keep a stallion at stud for use on his or her own mares, unless the stallion is of sufficiently high quality to render the progeny worth breeding, as there is seldom a market these days for inferior stock.

If however, you do decide to keep a stallion of your own he should be treated as much like other horses as possible. If he is not actually running out with his mares, he should be housed in a light roomy loose-box, with a half door so that he can look out and take an active interest in his surroundings. Boredom in the stallion often leads to bad temper and self-abuse. Wherever possible the stabled stallion should be turned out into a well fenced paddock for a few hours every day during the covering season and all day and night during the warmer weather out of the season – providing his paddock is very well fenced.

Prior to and during the season the stabled stallion must also be exercised on the lunge and then in-hand for increasing periods up to two hours each day or ridden out for an hour daily until he is fit. As with the hunter being got fit for the hunting season, the stallion should be got fit for the covering season. This means gradually increasing the amount of food and exercise the stallion gets, up to the beginning of the stud season and then regulating the food according to the number of mares he serves. A working stallion should not be allowed to get too fat otherwise his fertility may be affected, a good sign that a stallion is fit enough for his work, is that he does not sweat-up and blow hard when he is covering his mares.

It is unwise to use a two year old colt, as stud duties can tend to stunt his growth, but to a lesser extent than it does when a two year old filly is got in-foal.

For the first season at stud it is probably better to confine the stallion to 12 or 15 mares, this should not over tax him and should therefore give the mares every opportunity of getting in-foal. After the first season the number of mares can be gradually increased until 40 or 50 mares are taken. It is unwise to let the stallion serve any more than 50 mares in a season, as with increasing numbers his fertility will tend to drop sharply, unless both the stallion and his mares are very fertile. This is probably due in the main to the fact that although you may spread your mares out well over the season, if too many are covered at any one time, some of these will almost certainly break, giving possibly, double the quantity the following time. Should this ever occur it is then far better to leave some of the mares uncovered, concentrate on a few and get these in-foal, rather than cover all the mares and have them all return again; the unserved mares can then go to the stallion next time they come on, with an increased chance of conception.

The fertility percentage of a stallion not running with his mares depends to a very large extent on the thoroughness with which his mares are tried. Ideally all mares should be tried individually at least every other day; large public studs often try their mares by bringing the teaser (an inferior stallion kept for the sole purpose of trying mares) up to the paddock gate and observing the reaction of the mares in the paddock to his presence, this is a very labour-saving method of trying mares but has a major drawback in that shy maiden mares may never come up to the horse and will therefore be missed when they come in-season. Far better results are obtained by trying mares individually. Some mares show very little when they are in-season, sometimes only ceasing to kick violently, while others will always show to the horse, the only change being, that when they are not in-season they will swish their tails from side to side and lay back their ears. Thus careful observation of the mares' behaviour at all times is absolutely essential.

Many mares especially maiden mares will swing away from the trying bar thus making the process of trying doubly difficult. A mare should never be tried over a farm gate, as should she kick out, she might put her leg through the gate between the bars and break a bone. In the author's experience, a trying bar constructed as in the diagram on page 138 is the most effective.

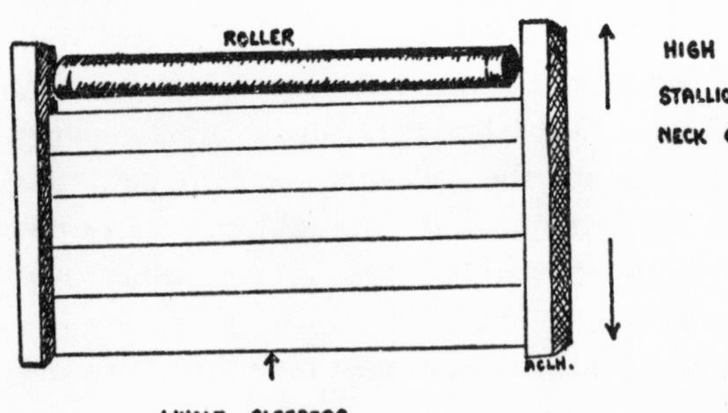

The stallion should be held back from the bar until the mare is in position. A recess should be provided near the mare's head to allow the man holding the mare to step back out of harms way, should the mare rear up and strike out, as sometimes happens.

When trying mares the mare and stallion should be allowed to touch noses and talk to one another, a fair idea of whether the mare is in-season or not will then be obtained, if there is no violent reaction to the stallion's advances, the mare can then be walked on a step or two and the stallion allowed to nip her flanks and talk to her. The stallion must be disciplined at all times, otherwise he may try to climb over the bar to get to the mare, for this season he must always wear a bridle when trying mares.

To obtain the best possible chance of conception the stallion should be confined to two mares per day or at the most three mares, as any greater number will reduce the chance of a fertile mating for the fourth and subsequent mares. Most fit stallions can cope with two mares per day, every day throughout the season and still maintain a high fertility but any greater number served every day would certainly tend to lower the fertility percentage. As already discussed in Chapter III, a mare should be served as near to her time of ovulation as possible to obtain the best possible chance of conception.

It cannot be stressed too much that the stallion's sheath must be kept clean, particular attention should be paid to the folds of skin at the top, which can become very dirty and greasy. The penis should be kept soft by applying oil or preferably Savlon

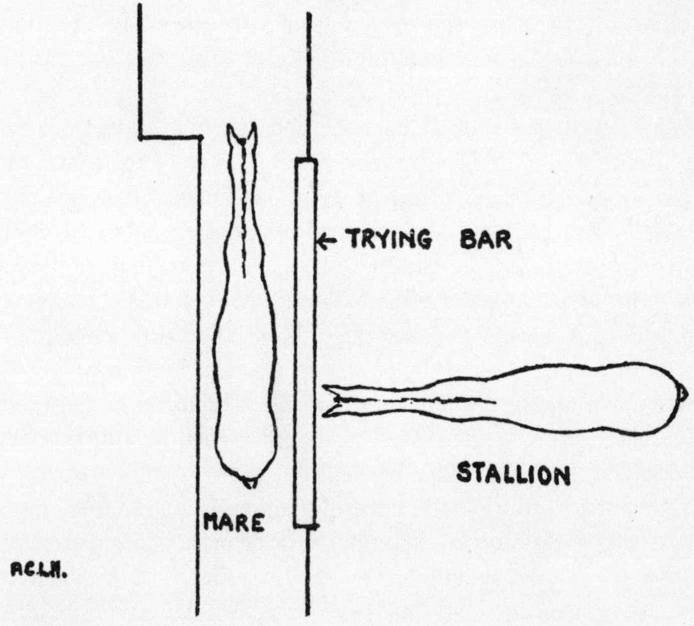

cream, which will both disinfect and soften the skin. It has no effect on the sperm and can be used immediately before service.

Ideally every barren and barren-maiden mare should be swab tested before service (when they are in-season) to make sure that they are free from disease, otherwise the stallion might become infected and become sterile, as well as passing on his infection to other clean mares.

The mare should be equipped for service as in the diagram on page 141.

The selected place for covering must be out of sight of your neighbours, otherwise you will receive complaints – and rightly so! It must have a non-slip surface such as sand, cinders, gravel or grass – but not mud, as the stallion can easily slip up should the mare move quickly once he is mounted. Preferably there should also be a slope or mound, either of natural or artificial origin, so that small mares can be placed uphill when covered by large stallions and big mares downhill to small stallions. There should be as little dust about at the time of service as possible, to cut down the risk of fungal infections in the mare, therefore covering outside is preferable to using a straw yard or barn.

Where a young stallion is concerned the first mare he covers should be chosen with care, she should be fully in-season and as far as possible she should be an older barren mare which is known to be very easy and quiet to cover, as a bad kick at the beginning of his career could upset a young stallion and make him timid or just disinterested. When a young colt is encountered which does not seem to know his job, he should be encouraged to mount a very quiet, fully in-season mare, if he fails to 'get the message', he should be put back into his box for an hour or two, after which time he should be tried again when he will probably have solved the problem for himself.

The normal routine for covering mares is: wash the mare's vulva with a disinfectant solution and clean her hindquarters if she is dirty or muddy and get her ready as shown in the diagram, leave your assistant holding her with the twitch on. Collect

TAIL
BANDAGE
SERVICE
BOOTS
A.C.L.H.
BRIDLE
TWITCH

your stallion which must be wearing a stallion bridle and long leather lead rein attached to the bit. Approach the mare on her near side at an angle so that she can see the stallion coming, keeping the stallion on her near side, either hold him back from the mare until he is fully drawn or let him approach the mare and try her but do not allow him to mount until he is ready. Always train your stallion to mount his mares from the near side and work his way round behind the mare, in this way he will be out of harm's way should the mare kick out as he mounts.

A little vaseline smeared on the mare's vulva can sometimes help the stallion but in any case the stallion's penis should be guided gently into the vagina, at the same time the mare's tail must be pulled to one side to prevent any tail hairs from being taken into the vagina with the penis, as should this happen the penis might be cut.

As the stallion serves his mare, it will be observed that his tail 'flags' i.e. waves up and down, this happens to a greater or lesser extent according to the individual stallion's conformation, being most noticeable in the Arab and others with a high tail carriage. Apart from watching the flagging of the stallion's tail, a hand can be placed gently on the underside of the penis and the pulsations felt, in this way the effectiveness of the service can be judged. When the stallion has finished serving his mare, he will usually release his hold on her and relax but he should not be forcefully pulled off his mare but allowed to rest on her should he so wish. The moment he comes out of the mare some disinfectant solution (which can be obtained from most horse specialist veterinary surgeons) should be thrown over his penis and hind legs. As he dismounts from the mare her head must be pulled round towards the stallion leader, thus automatically turning her hindquarters away from the stallion so should she kick out as he dismounts, she will kick past the horse and not injure him.

The stallion should be returned to his box and his nostrils and chest washed down with a disinfectant solution, to kill the smell of the mare in-season. If the stallion is required to serve any further mares that day, he should be allowed at least four hours rest between coverings.

Once the mare has been served the service boots should be unfastened and the mare allowed to step out of them, if she shows any signs of straining she should be walked out for about 10 to 15 minutes and on no account allowed to stop. Otherwise she should be turned straight out into the paddock.

Maiden mares often have a skin – Hymen – either wholly or partially across the vagina which must be broken before an effective service can take place. This may be done naturally by the stallion when he serves the mare, a little blood will be noticed to escape from the vulva after service, in which case it is better to cover the mare again the following day, especially if she is just going out of season.

When a small stallion covers a large mare he can be assisted by holding his front legs on either side of the mare to support him. A rope can be tied round the mare's neck and anchored to the mane, which the stallion holds with his teeth. Some stallions tend to savage their mares, biting their necks badly, in which case a leather guard should be placed on the mare to protect her.

Some studs use hobblers when serving mares but in the author's experience these tend to upset mares, especially those of Thoroughbred or near thoroughbred breeding, whereas most mares, accept service boots without much fuss: these can be obtained from most saddlers. Occasionally a really bad mare is encountered but in many cases on examination by a veterinary surgeon it will be found that the mare is not properly in season, but should she be found to have a ripe egg an injection of tranquilliser is preferable to a fight and certainly has no ill effects and might prevent a lamed stallion.

In nature the mare tends to walk on a step or two as the stallion mounts her and when she is held for service she should still be allowed to walk on, unless she is too big for the stallion and is being held on a steep slope when she can be held still BUT ON NO ACCOUNT SHOULD A MARE EVER BE

PUSHED BACK A STEP otherwise the stallion may lose his balance and fall over backwards.

An accurate record must be kept of each mare during the covering season, a useful record card for this purpose is shown on page 145: Wall charts may be obtained, but these are of greater value to the commercial stud than to the amateur horse breeder.

Where mares are being covered for sale a record of the mares' foaling returns can be compiled by sending out cards at the end of the foaling season, for the owners to complete, this is of particular interest where non-pedigree stock are concerned and no official foaling returns are made. Records of the stallion's progeny in the show ring and on the race course each year can also be kept.

According to the Horse Breeding Act 1958 it is illegal to keep a colt or stallion over the age of two years without a licence or permit except for certain exceptions which are:

1. Thoroughbreds – i.e. horses entered or eligible for entry in the General Stud Book.

2. Certain breeds of pony turned out or standing in an area where ponies of his breed usually run on free range, providing he is not travelled or exhibited for service, i.e. Dartmoor, Exmoor, New Forest, Welsh Mountain, etc.

A licence or permit is required as soon as the colt is two years old – his age being calculated from January 1st of the year he was born. Application for a licence should be made to: Livestock Improvement Branch, Ministry of Agriculture, Fisheries and Food, Great Westminster House, Horseferry Road, London, S.W.1, or in Scotland to Broomhouse Drive, Edinburgh 11, between 1st July and 30th September of the colt's yearling year. The current licensing fee is eight guineas per stallion which is not returnable if the licence is refused.

Before a licence is issued the animal will be examined by a veterinary surgeon appointed by the Ministry and may be re-examined from time to time to ascertain that the licence should not be revoked. A licence may be refused or revoked on any of the following grounds:

1. A contagious or infectious disease.
2. Cataract, roaring, whistling, ringbone (high or low), sidebone, bone spavin, navicular disease, shivering, stringhalt, or defective genital organs.
3. Inadequate fertility.
4. Defective or inferior conformation or physique.

A stallion over nine years old will not have his licence revoked for defective wind if the licence has been in force for at least two years.

If the stallion fails on inspection, the owner may appeal on payment of 13 guineas, which will be returned in the event of the appeal being turned down. In this case the stallion will be examined by a referee; if he is turned down the owner will be instructed to have him castrated or slaughtered.

A permit may be issued when the Ministry decide that for some reason a licence should be temporarily refused but that the stallion should be kept entire. It is usually a condition that the stallion should not be used at stud until the full licence has been granted.

A stallion owner may be asked to produce his licence to a police officer, Ministry official or person in charge of a visiting mare and it is an offence if he refuses.

When a stallion is sold or leased for longer than six months the licence should be sent to the Ministry for transfer to the new owner. Should the stallion remain in the same ownership but be moved to a new address the Ministry should also be notified.

No. ....   DESCRIPTION OF MARE   Served by..................................

Colour ....   Breed... ....   Name ....

IF FOAL WAS BORN ALIVE			If slipped give date	If born dead give date	If barren write barren	If mare died before foaling was she believed in foal or barren
Date of Birth	Colt or Filly	Colour				

If mare was sold before foaling, please give name and address of person to whom sold.

....

....

Signature of owner of mare ....

Address ....

....

INDIVIDUAL MARE RECORD CARD:

NAME ______________________

BREEDING ______________________ AGE ______________

	1	2	3	4	5	6	7	8	9	10	11	12	13	14	15	16	17	18	19	20	21	22	23	24	25	26	27	28	29	30	31
FEB.																															
MAR.																															
APR.																															
MAY.																															
JUNE.																															
JULY.																															

X : IN SEASON
S : SERVED
W : WORMED
F : FOALED
■ : TESTED IN-FOAL
O : TESTED NOT IN-FOAL
B : BLACKSMITH

STALLION "X"

1970 FLAT RECORD OF PROGENY

AGE	NAME	SEX	DAM	SIRE OF DAM				
2	BILLY	C.	SUSIE	FAIR TRIAL	KEMPTON - APR. JUNIOR STKS. 5fur. £345. 2nd.	NEWMARKET - MAY. FEN DITTON STKS. 5fur. £650 1st		
2	SPOTS	C.	HIGH GIRL	VULGAN	WARWICK - MAR. WOODEND STKS. 5fur. £360. (7th-10)	SANDOWN - APR. SPRING JUV. STKS. 5fur. £630. (10th-20)		
3	DUSTY	f.	FLYER	DENTURIUS	NEWMARKET - MAR. HEATH H'cap. - 1m - £850. (5th-12)			
3	RUBBER	C	ALWAYS	HYPERION	YARMOUTH - APR. SEASIDE S. STKS. - 1m. 2f. - £260. 3rd.	ALEXANDRA PK. - MAY. ROSE S. STKS. - 1m. - £345. 1st.	EPSOM - JUNE DOWNS S. STKS. 1m. 4f. £345. (6th-12)	
3	BARREL	f	GUNNER	COMBAT	TEESIDE - MAY. WATER H'cap. 1m. 4f. £670. 2nd.	CATTERICK - JUNE CAMP H'cap. 1m. 2f. £455. 1st.	DONCASTER - JUNE. SALES H'cap. 1m. 4f. £770. (8th-20)	RIPON - JULY CATHEDRAL H'cap. 1m. 4f. £470 2nd.
4	PAPER	g	PRINTER	BRIGHT NEWS	HAYDOCK - JUNE MG. H'cap. 1m. £477. 1st.	WARWICK - JUNE CASTLE STKS. 1m. 2f. £1350 3rd.	WOLVERHAMPTON - JULY MIDLAND STKS. 1m. £850 1st	

14
THE ORPHAN FOAL

An orphan foal is strictly speaking a foal which has lost its mother, fortunately a very rare occurrence, but other foals may be deprived of their mother's milk due to a variety of reasons, e.g. the mare rejecting her foal at birth, foaling down without any milk or becoming ill at some stage during the lactation period and drying up. Fortunately all these cases are very rare. However, should anything like this happen you will have two courses open to you:

i) *Rearing the foal by hand:*

As we have already seen in chapter 5 page 51 the foal *must have colostrum within the first* 36 *hours of birth,* in order that it can develop resistance to most of the diseases with which it will come in contact. Many of the large equine veterinary practices keep a supply of mares' colostrum in deep freeze for emergency use, but where it is impossible to get hold of any colostrum *it is essential* that you should call in your veterinary surgeon so that he can give the foal some substitute colostrum and antibiotic or serum injections.

All foals must have some form of milk in order to survive. For the first four weeks the foal must be fed every two hours both day and night; then the length of time between feeds can be gradually increased. Foals cannot be treated like calves, but must be fed little and often: over-feeding at any one time will make a foal ill.

At first, the foal should be fed from a bottle fitted with a rubber 'lamb teat', obtainable from most agricultural chemists and measuring 4 in. x 1 in. Milk is also an ideal food for bacteria (germs), so any bottles, teats and mixing utensils used for feeding

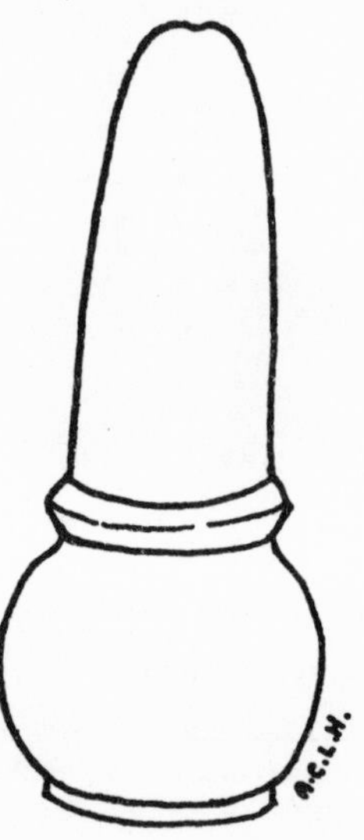

TEAT FOR FEEDING FOALS
(4″ x 1″)

must be kept absolutely clean and sterilized between each feed. Sterilization can be carried out with boiling water or by using a dairy chemical sterilizer such as hypochlorite, in which case care should be taken to make sure that the surface is absolutely clean before treatment. If the equipment is not sterilized regularly the foal will probably start to scour and become quite ill. To avoid digestive upsets the milk should always be fed at the same temperature, preferably blood heat. For this reason, it is often better for the same person to feed the foal all the time wherever possible. Unused milk should be thrown away and never re-warmed.

A satisfactory substitute for mares' milk can be made up easily in the following way:

2 tablespoonfuls of lime-water (obtainable from any chemist)
4 tablespoonfuls of glucose
made up to 1 pint with warm cow's milk.

Goat's milk, when it can be obtained, will give better results than cow's milk and should be fed as a direct substitute for cow's milk in the above recipe.

Where cow's milk is difficult to obtain in sufficient quantity the foal can be given Ostermilk Number 1 at the rate of:

From birth to approx. 1 month or according to size of foal	To weaning
per feed	*per feed*
4 measures of Ostermilk	10 measures of Ostermilk
2 measures of glucose	4 measures of glucose
to 8 fl. oz. of water	1 pint of water

The Ostermilk should be mixed to a paste with a little cold water until there are no lumps left, then hot water added and the temperature adjusted to blood heat. Too much liquid will cause the foal to develop a 'pot belly'.

The feed should measure from 8-20 fl. oz. or more according to the size of foal. Ostermilk (which is really designed for human babies) is a rather expensive method of rearing foals by hand through to weaning. A specialist milk powder which has been formulated exactly to the requirements of orphan foals is now on the market under the trade name of 'Equilac' and can be obtained from the manufacturers or through the National Foaling Bank (see page 153). This contains all the vitamins, minerals and trace elements necessary for the young foal in a high quality milk powder composition, and should be fed according to the makers' instructions:

Age	*Equilac*	*Water*	*Pegus Junior*
0-2 days	Colostrum (Biestings)		
3rd day	1 lb.	4 pts.	
4th day	1½ lb.	6 pts.	
5-7 days	2 lb.	8 pts.	
2-4 weeks	3 lb.	12 pts.	½ lb.
5-9 weeks	4 lb.	12 pts.	½ lb.

10-12 weeks	4 lb.	10 pts.	2 lb.
13-14 weeks	2 lb.	5 pts.	4 lb.
15-16 weeks	1½ lb.	4 pts.	6 lb.

Most foals are born with a well-developed suck reflex and those which are slow in this respect will usually develop the desire to suck within two hours of getting to their feet. Ideally the foal should have its first feed within two hours of birth and certainly not later than four hours. Mix up the milk as described above and put it in the feeding bottle. If you have no milk or equipment, it is better to wait until the shops open in the morning than risk upsetting the foal with a home-made substitute. Be careful not to over-feed.

Make sure the hole in the rubber teat is large enough to allow a good flow of milk to escape. Give the foal a taste of the milk by squirting a little into its mouth, this should awaken its suck reflex. Make sure your hands are clean, then you can run a little milk onto your fingers and get the foal to suck them and gradually substitute the teat for your fingers. Unless he is lying down, it is better to get the foal's hindquarters into a corner when you start getting him used to sucking, so that you will have some control over him. If the foal does not start sucking soon, squeeze some milk out of the end of the teat into its mouth, until he finally gets the idea. When the foal is hungry he will soon start sucking without much trouble.

The firm which manufactures Equilac, also makes a Junior nut which they recommend for feeding with the milk substitute. It is best if you can get the foal to eat some form of solid food as soon as possible. Most foals will nibble a little warm bran mash from your fingers the day after they are born; something warm and wet will encourage them more than a dry feed. Therefore a little bran, linseed, rolled oats and flaked maize dampened with some milk is highly suitable. Once the foal is beginning to eat he should be introduced to the Junior nuts, which it is suggested should be fed at the rate of 1 lb. per day per month of age.

I have found that ad lib feeding of nuts is possibly to be preferred, as the foal can then help itself whenever it wishes. A boiled barley and linseed mash containing a raw egg is best fed every night, to give some variety to the diet. The foal must also receive ad lib hay and water.

A foal which is being reared by hand should have some form of companion, otherwise he will tend to become very lonely, and will often not 'do' well. A quiet sheep, goat or big dog is suitable. Make sure the foal is kept warm; in cold weather an Infra-Red lamp may be necessary.

ii) Get a foster mother:

Most hand reared foals become very 'humanised' and seldom have much respect for their owners. It is therefore desirable to bring any orphan foal up on a foster mare if at all possible.

If you intend to find a foster mother for your orphan foal, do not teach the foal to drink out of a bowl as it is then more difficult to persuade it to suckle a mare at a later date. If you do wish to teach the foal to drink out of a bowl or bucket which saves time, the easiest way to do this is to get him to suck your fingers and then gradually lower your hand into the milk in exactly the same way as one would with a calf. A little food scattered in the bottom of the bowl when the foal has nearly finished drinking will encourage him to eat.

Rearing a foal by hand is a tedious but rewarding job, necessitating getting up every two hours, throughout the night at least for the first month. The problem of finding a suitable foster mother, without any help, is usually insurmountable but there is an organisation in this country known as the National Foaling Bank which is geared to help anyone who has either lost a mare or foal. The address of the chief organiser is:

Miss Johanna Vardon,
Meretown Stud, Newport, Shropshire.
Telephone Number: Newport (Salop.) 811234 (STD code 0952).

APPENDIX

There are many instances when you may require to take a sample from your horse to give to your veterinary surgeon for testing. In these cases it is very useful to know the amount of sample required by the laboratory for an accurate result.

Only samples which are usually collected by the owner for the veterinary surgeon are listed and not those collected by the veterinary surgeon himself.

Faeces samples:

1. Worm egg count – for Redworm, Roundworm, Tapeworm or Liver Fluke:

   Enough sample to fill an Equizole powder tin or universal bottle.

2. Lungworm larvae count:

   The minimum quantity required for a satisfactory determination is a half pound honey jar absolutely full; a less amount gives an inaccurate result.

3. Occult blood test:

   Minimum quantity is a universal bottle full, the container must be sterile.

4. Bacteriology:

   Minimum quantity, as above but collected into a sterile receptical; it must be a really fresh sample, preferably taken before it reaches the ground.

As far as faeces samples for worm counts are concerned, they should be put into an air-tight container such as a sealed tin, wax pot or polythene bag – they must not be allowed to dry out, they should therefore be placed somewhere cool and not in direct sunlight or on a radiator, otherwise the eggs will tend to hatch out and a false result will be obtained. All samples must be from absolutely fresh droppings.

Urine samples:

1. For pregnancy testing:

   A minimum quantity of a little under half a universal bottle full is quite sufficient but a larger quantity should be sent wherever possible. The container need not be sterile but must be absolutely watertight.

2. For routine examinations:

   The minimum quantity is a one pound honey jar full, the container should be absolutely clean and dry and should preferably be sterile.

3. For bacteriological examinations:

   The minimum quantity is a universal bottle full, the container must be sterile and the sample should be obtained straight into the bottle from mid-stream of the urination.

Skin scrapings:

These are usually taken for the evidence of mites, ringworm and dermatophalus – lice can be seen with the naked eye.

For these examinations the sample must be plucked or scraped from the actual site of the infection and incorporating as much skin tissue and hair roots as possible and not just the hairs themselves. On no account should the hairs be cut or clipped off.

Sufficient sample to fill a match box, if possible, would be ideal for the tests carried out.

As ringworm is very contagious for man, a pair of rubber gloves should be worn when collecting the sample.

Milk samples:

These are usually collected where inflammation and disease of the udder is suspected.

The sample must be collected by a sterile technique into a sterile container, minimum one universal bottle full. Using a piece of cotton wool soaked in methylated spirits swab your hands and the mare's teats, draw some milk from the affected quarter (teat) or quarters into the sterile bottle. Give the sample to your veterinary surgeon as soon as possible.

Foetus and membranes:

Should you have the misfortune to own a mare which slips her foal in one or more years and you happen to find the foetus and membranes, these should be collected into a clean polythene bag and given to your veterinary surgeon as soon as possible. On no account should the foetus and membranes be washed, not even with plain water let alone disinfectant! Wherever possible make sure that both the foetus and membranes are included.

A universal bottle is a small glass bottle with a metal screw top and measures 8.5 cm. high, and has a diameter of 2.75 cm.

DATE COVERED

JULY — AUG. — SEPT. — OCT. — NOV. — DEC.

1 11 21 31 10 20 1 9 19 29 9 19 29 8 18 28 8 18 28

6 16 26 6 16 26 5 15 25 4 14 24 4 14 24 3 13 23 3

JUNE — JULY — AUG. — SEPT. — OCT. — NOV.

DATE DUE

DATE DUE

DEC. — JAN. — FEB. — MAR. — APR. — MAY

7 17 27 6 16 26 5 15 25 7 17 27 6 16 26 6 16 26 5

1 11 21 31 10 20 2 12 22 1 11 21 1 11 21 31 10 20 30

JAN. — FEB. — MAR. — APR. — MAY — JUNE

DATE COVERED

INDEX